WILD RITUALS

WILD RITUALS

10 LESSONS ANIMALS CAN TEACH US ABOUT CONNECTION, COMMUNITY, AND OURSELVES

CAITLIN O'CONNELL, PhD

CHRONICLE PRISM

Copyright © 2021 by Caitlin O'Connell.
Photographs copyright © 2021 by Caitlin O'Connell & Timothy Rodwell.
Wolf images courtesy of Jim and Jamie Dutcher, *Living with Wolves*.
Flamingo image courtesy of Melissa Groo.
Chimpanzee image courtesy of Frans de Waal.
Humpback whale and calf image courtesy of Rich Neely.
All rights reserved. No part of this book may be reproduced in any form without written permission from the publisher.

Library of Congress Cataloging-in-Publication Data
Names: O'Connell, Caitlin, 1965- author.
Title: Wild rituals : 10 lessons animals can teach us about connection, community, and ourselves / Caitlin O'Connell, PhD.
Description: 1st. | San Francisco, California : Chronicle Prism, [2021] | Includes bibliographical references. |
Identifiers: LCCN 2020038985 | ISBN 9781452184852 (hardcover) | ISBN
  9781797201634 (paperback) | ISBN 9781797201610 (ebook)
Subjects: LCSH: Animal behavior.
Classification: LCC QL751 .O28 2021 | DDC 591.5—dc23
LC record available at https://lccn.loc.gov/2020038985

Manufactured in the United States of America.

Design by Pamela Geismar.
Typesetting by Maureen Forys, Happenstance Type-O-Rama.
Typeset in Bauer Bodoni, Brandon Grotesque, Baskerville.

10 9 8 7 6 5 4 3 2

Chronicle books and gifts are available at special quantity discounts to corporations, professional associations, literacy programs, and other organizations. For details and discount information, please contact our premiums department at corporatesales@chroniclebooks.com or at 1-800-759-0190.

CHRONICLE PRISM

Chronicle Prism is an imprint of Chronicle Books LLC, 680 Second Street, San Francisco, California 94107
www.chronicleprism.com

Thanks, Dad, for showing me my first crayfish and newt in the stream in the woods behind our house in Wyckoff, New Jersey, for showing me how to fish at Culver Lake, for taking me on camping trips, and for teaching me how to scuba dive when I was nine years old. Thanks for sharing your passion for nature and wanderlust for travel—clearly, it was contagious.

CONTENTS

INTRODUCTION
The Lost Art of Ritual · *1*

1 GREETING RITUALS
Spit, Snot, and Other Social Grease · *17*

2 GROUP RITUALS
The Power of the Collective · *31*

3 COURTSHIP RITUALS
Outlandish Attractions · *47*

4 GIFTING RITUALS
Shiny Objects, Flowers, and Dead Birds · *63*

5 SPOKEN RITUALS
Rumbling, Roaring, and Yodeling · *77*

6 UNSPOKEN RITUALS
The Gravity of Proximity, Postures, and Expression · *97*

7 PLAY RITUALS
Catch a Lion by Its Tail · *117*

8 GRIEVING RITUALS
Ritualized Healing through Time · *133*

9 RITUALS OF RENEWAL
Spring Cleaning and Other Rhythms in Nature · *151*

10 RITUALS OF TRAVEL & MIGRATION
Forest Bathing and a Journey to Oneself · *165*

NOTES · *185*
ACKNOWLEDGMENTS · *219*
ABOUT THE AUTHOR · *223*

INTRODUCTION

THE LOST ART OF RITUAL

> "The love for all living creatures is
> the most noble attribute of man."
> —CHARLES DARWIN

HAD A LOT ON MY mind while driving back to Mushara waterhole—my elephant field site in the northeast corner of Etosha National Park, Namibia. It's a place I've returned to every July, for the last thirty years, to study elephants. One of the largest national parks in Africa, Etosha spans over 8,500 square miles and hosts approximately three thousand elephants.

Lost in thought, my eyes focused on the dusty horizon. Suddenly, two gray behemoths appeared in the middle of the chalky calcrete road, oblivious to the vehicle hurtling toward them.

I slammed on the brakes and pulled over to avoid colliding with two of the world's largest land creatures. Right in front of me, two female elephants hell-bent on reuniting were kicking up a large cloud of white dust to get to each other.

I was already running an hour late and I was eager to get back to the field site before what I refer to as "elephant o'clock"—when family groups of elephants start convening at waterholes. This window is anywhere between four o'clock in the afternoon until about two o'clock in the morning. If my research team and I had any hope of building our elephant identification catalog, we had to take photos before the sun went down.

When the dust settled, I could see these giants were two of my favorite elephants, Knob Nose and Donut, who were named for their distinctive physical features. The former had a very large wart on her trunk. The latter had a very large hole in the middle of her ear like a donut hole. They had come from opposite sides of the road, and upon seeing each other, immediately ran to embrace, their ritualized greeting ceremony blocking my passage.

Facing each other, the elephants held their heads high above their shoulders while rapidly flapping their ears. Then Donut lifted her trunk and bellowed a thunderous roar, almost as if something terrible had just happened. Having observed wild elephants for as long as I had, I knew this vocalization was purely an expression of intense excitement.

Next, the elephants rumbled softly, while preparing to place their trunks in the other's mouth—the elephant equivalent of a handshake. Trunks extended, the tips of both of their trunks quivered in anticipation, as Donut gently placed the tip of her trunk against the side of Knob Nose's mouth like a kiss. Knob Nose reciprocated.

After the requisite trunk-to-mouth greeting ritual, they immediately positioned themselves side by side, facing north. They stood with a foot-long section of their trunks lying flaccid on the road. When an elephant does this, it appears that they've suddenly lost all muscle control over their enormous, prehensile noses. Their shoulders were erect and poised, as if they were about to engage in a formal march. Instead, they remained frozen in place, while roaring and rumbling wildly.

Then the inevitable happened. No female elephant greeting ceremony is complete without the sudden and thorough evacuation of both bowels and bladder. It is the ultimate expression of sheer, elephantine joy.

introduction

Given the intensity of their greeting, you might assume they hadn't seen each other in years. While I had no way of knowing exactly how long they had been separated, my guess was minutes, to perhaps a few hours. Whenever I'd see one of these elephants, the other wasn't far behind, so I couldn't imagine it had been very long.

Knob Nose and Donut were residents in the region of my elephant study, Knob Nose being the matriarch of her family, and Donut, second in command. Since they spent much of their time at a neighboring waterhole, I didn't get to observe them very often. I had always assumed that Donut was Knob Nose's daughter—such was the age difference and the nature of the bond they shared.

Knob Nose and Donut continued vocalizing in long low rumbles, ears still flapping rapidly back and forth, contributing to the frenetic mood of the ritual. The temporal gland next to their eyes was now streaming, creating two wet streaks down either cheek.

Clearly something momentous was occurring during their reconnection, both psychologically and physiologically. The intent and singular focus with which they engaged in this very specific set of ritualized gestures was striking.

THESE TYPES OF RITUALS in the animal kingdom may seem completely unrelated to rituals in our own lives. But they aren't and shouldn't be. Greeting rituals are pivotal to peaceful coexistence. Observing how important greeting rituals are in other animals is an important reminder of our own need for this ritual.

Even seemingly small acts in our daily lives—like saying hello, bowing, making eye contact, or giving someone a hug—are things we sometimes take for granted. Rituals related to greetings, courtship, bonding, play, and mourning, for example, are a huge part of our own lives, and much is lost when we ignore them. Rituals inform our behavior, particularly when we are feeling uncertain about what to do.

They provide a routine in an unpredictable world. They also hold us together as a community with a shared set of expectations. We have much to learn from our nonhuman animal relatives.

In fact, we have a tremendous amount in common with other animals—and plants for that matter. We even share 50 percent of our genes with a banana. I was startled to learn this fact one afternoon during a visit to the Museum of Natural History in Washington, DC, while exploring the *Homo erectus* exhibit and its graphs explaining human ancestry. I was aware that we share 61 percent of our genes with the fruit fly and 85 percent with the mouse—and even 98 percent with our nearest relative, the chimpanzee. But *a banana*? Not only does a banana not have a brain or a spine, it isn't even an animal.

Many of the genes we share with the banana are called "housekeeping" genes, which are necessary for basic cellular functions such as breathing, repair, and replication. Since both plants and animals need to consume either carbon dioxide or oxygen as well as reproduce to survive, it's no wonder that all life shares some basic coding. This museum exhibit reminded me that we are interconnected with all other organic things on this planet.

Recent genetic findings point to all current life on Earth evolving from a single-celled organism that originated approximately three and a half billion years ago. This organism, named LUCA, an acronym for "last universal common ancestor," was our humble beginning on this planet.

Over the last few decades, the scientific community has debated whether life began in an extreme environment, such as high salinity or temperature—in deep-sea vents or near volcanoes—or in a "warm little pond" with access to photosynthesis, as Darwin envisioned. Now, we can all take comfort in knowing that our lineage most likely originated in a deep-sea vent in the Galapagos.

Single-celled organisms needed about three billion years to evolve into multicellular organisms, and all multicellular organisms (including humans and bananas) share a common ancestry that goes back less than a billion years. This is how humans ended up with genetic similarities to a tropical yellow fruit.

If you still have a hard time conceptualizing this, ask yourself why humans have gill slits and a tail as embryos. Gill slits are found in the embryos of all vertebrates because all vertebrates share a fish from four hundred million years ago as a common ancestor. Eight million years ago, mammals as diverse as horses, tigers, whales, bats, and humans all shared a common ancestor in the form of a tiny shrew-like creature, which is why mammals share such defining features as mammary glands, body hair (or fur), and three middle ear bones.

As human beings, we often try to distinguish ourselves from other animals as being more advanced—or superior. Recognizing our similarities to other animals can instead be cause for celebration, not treated like a threat to a misguided sense of uniqueness, separateness, or even dominion over the rest of nature.

This is particularly true of rituals. Animals have incredibly sophisticated rituals that are related to all aspects of their lives. These rituals allow them to survive in a very complex world, to predict what will happen next, and to connect deeply with their families and communities. These rituals are very similar to our own.

Modern brain-imaging technology helps us understand what goes on inside the minds of other social animals. For instance, by comparing how the brain works in human and nonhuman primates—along with many other animals, including dogs and even reef squid—scientists have shown that we have similar ways of using our brains. In addition, the same hormones are expressed under similar psychological and social circumstances. Other studies have shown that many animals also experience a lot of the same emotions we do.

I am continually amazed by how much we can learn about ourselves from the wild animals that have captivated our imaginations throughout our evolutionary history. Every time I witness the intimacy of an elephant greeting ritual, or a moment of cooperation between an elephant grandmother and her daughter to save the mother's calf from danger, I am reminded of how similar our societies are. When a young male elephant chews food for an elder that no longer has his teeth, how can we not be moved by this act of kindness? It's impossible not to see the similarities in our own caregiving to elders.

I can't help thinking that if we could accept emotionally intelligent animals as strange and wonderful extensions of ourselves, we might be more compassionate toward other animal societies. After all, we share a common ancestor. In turn, this may also inspire us to embrace different societies within our own species with a more generous spirit. This book is an exploration of shared rituals that occur within both wild animal and human societies, and it offers a path to accepting our similarities—and ultimately, to appreciating our differences.

CONSIDER A LONE male chimpanzee approaching a large fig tree in the middle of a dappled forest in Côte d'Ivoire. He sits down and stares at the tree intently. He looks away for a moment, scratches his arm, and then refocuses his attention on the tree.

Suddenly, the chimpanzee stands and picks up a rock the size of a melon. His shoulders start heaving and his lips purse as he emits a soft moaning sound. The moaning gets louder and louder, shorter and sharper in a buildup that reaches an open-mouthed climax—the chimpanzee's signature pant-hoot vocalization.

At the height of this intense call, the chimpanzee reveals his intention and hurtles the rock against one of the buttresses of the tree with a bang. He then climbs onto the buttress and briefly beats at it with his feet as if he were playing a drum, a behavior known as "drumming." After a good drumming, he runs off into the forest, screaming.

This strange ritual is called "accumulative rock throwing," a behavior that has only been documented in four different chimpanzee populations within West Africa. At all four rock-throwing ritual sites, researchers have found an accumulation of rocks in front of specific trees, each with fresh scars on their bark. The selected trees had either buttresses or hollow trunks, likely chosen to create louder sounds. Over a period of four years, camera traps set up at these sites recorded sixty-three accumulative rock-throwing rituals, many instances involving the same individuals.

Researchers discovered the ritual was almost always carried out by adult males and always included the same three elements: picking up a rock, throwing the rock at the same tree, and using the pant-hoot vocalization. Researchers suggest that rock throwing could have evolved to enhance the drumming display, helping to project the sound of the ritual—perhaps to broadcast a territorial claim even further than drumming alone.

Drumming on its own is a well-known ritual across chimpanzee populations; it is used to define territories or obtain a potential mate. The drumming produces a loud low-frequency sound that travels over half a mile away. Some particularly fascinating studies have shown that drumming patterns in chimpanzees can convey the identity of the individual drummer. Some suggest that these patterns could even be a precursor to musical rhythm.

Chimpanzee rituals are thought to be homologous to our own, evolving from a shared ancestor rather than evolving in parallel. Researchers believe they serve as models for how our early hominid ancestors may have behaved and shaped rituals around behaviors like hunting, demarking territories, and designating a ritual site.

Chimpanzees enact a ritual dance at the onset of rain and sometimes when coming upon a waterfall. Primatologist Jane Goodall suggests that chimpanzee displays centered on natural elements may be precursors of religious ritual.

Chimpanzee accumulative rock throwing shares several important features with human ritual practices. The first is a strong association with a particular site, like the tree. The second is the accumulation of artifacts left over time, such as the pile of rocks. Last, there is a specific set of ritualized behaviors—picking up a rock, throwing the rock at a designated tree, and emitting a specific vocalization.

The accumulative rock-throwing sites could serve as a territorial boundary or navigational markers for pathways, just as rock cairns have done for many human societies (and still do for hikers). Stone-accumulation shrines at sacred trees are known among indigenous West African peoples, and ritual sites like these are thought to have led to the foundation of religion.

Humans have been engaging in rituals for tens of thousands of years. Archaeologists recently found the oldest fossil evidence of a sacred ritual site at Tsodilo Hills in Botswana, where the San people worshipped the python, dating back seventy thousand years.

The python is the San people's most important animal. In their creation myth, humans descended from the python, and the dry streambeds circling the Tsodilo Hills were created by the python in search of water. Archaeologists have found a specific ritual site in a small cave in the hills with artifacts and paintings and a large rock carved to look like a python.

Arrowheads found at the site were made of precious stone that had to have been brought from hundreds of miles away. Anthropologists believe a secret chamber in the cave housed the shaman, the religious leader, representing the python. There was no evidence of habitation. The site was purely used for ritual.

Africa is not just the place where humans were born. This ritual site at Tsodilo Hills is the first evidence that early *Homo sapiens* engaged in abstract thought related to ritual, long before modern cultural practices appeared in Europe.

Ritual is often thought of solely in the context of a religious ceremony. However, rituals span far beyond religion, cult, or spiritual practices. A ritual is a specific act or series of acts that are performed in a precise manner and repeated often. It can be as simple as your daily practice of "saluting the sun" in yoga or as complicated as playing Beethoven's Symphony No. 5 on the violin for the New York Philharmonic on Friday evenings. Ritualized actions are typically exaggerations of otherwise normal behaviors, like throwing a rock. While each action within the ritual in and of itself isn't always meaningful, *the total result is.*

Primatologists Tennie and van Schaik, offer a very narrow definition of ritual. They reason that all behaviors that are to be considered ritual must be learned not inherited, performed within a group setting, and evolve differences between populations in order to be considered true ritual. This definition rules out all non-human primate behaviors that other experts have deemed ritual, including accumulative rock throwing, grieving and even the suggestion of a prototypical trans-species definition of religion. For the purposes of this book, I use a much broader definition of ritual that links ritualized social behaviors, whether inherited or learned, across many species within many different contexts as defined by experts in their respective fields.

In order to understand the importance of ritual, we need to understand that each action is typically more impactful and multi-layered than it seems. Performing all the steps within a ritual in a specific sequence often requires complete focus to be achieved successfully. The science behind these acts shows how engaging in ritual can relieve stress, decrease anxiety, make us more present, and even improve our cognition.

When we exaggerate a familiar behavior in a ritual practice, it alerts our minds to an unusual stimulus that requires focus, activating areas of the brain like the amygdala—which is responsible for processing our emotions and responses. In addition to helping us tap into our emotions, repeating a sequence of ritualized steps can be critical for learning and long-term memory, allowing us to improve our concentration, problem solve faster, and think on a deeper level.

Rituals are exhibited across all human and nonhuman animal societies in some way or another. In their simplest form, they are a tool to communicate and express intentions. They also create a mutual language to facilitate connection—not unlike sports fans chanting a team slogan or cheer, which both sends an intimidating message to their rivals and acts as a unifying call that bonds the fans and rallies the team.

Anthropologists believe that rituals originating in early human societies often included an additional element that addressed the "hazard-precaution system." Behaviors such as washing and cleaning, or creating an orderly environment, were incorporated into rituals to address concerns of food, perimeter security, or healing in some way. Rituals that combine positive prescriptive elements (skip ten steps down the sidewalk) with negative ones (don't step on any cracks) force the mind to engage memory and motor control in a way that wouldn't occur with a regular routine.

Research shows engaging in ritual is temporarily appeasing and mitigates anxieties. For instance, even the simple act of marking the boundaries of a campsite, or drawing a circle in the sand around ourselves, can be relaxing. It marks our perimeter and gives us a sense of security.

During the performance of a group ritual, personal fears or doubts are shared as a group, which has a calming effect. Engaging in ritual also has a profound impact on the hormone expression of all participants, which results in physiological, immunological, and behavioral changes. Ultimately, this helps to create cooperative relationships within complex societies.

Whether simple or elaborate, rituals can be transformative, both mentally and physically, and they connect us, strengthen bonds, create order, and ground us within a community. They *are* the glue that binds communities together into healthy societies for all social animals. Since social isolation is a major risk factor for mortality, in both humans and nonhuman animals, ritual plays an important role in bringing us together and keeping us healthy.

An elephant trunk-to-mouth greeting ritual, for example, can be much more than just a greeting. Anyone witnessing this ritual for the first time quickly recognizes how trusting it is for an elephant to place the tip of its trunk in another's mouth. It's a risky behavior considering how sensitive the tip of an elephant's trunk is and how easy it would be to get bitten.

This trunk-to-mouth ritual is akin to a handshake and signals respect. An elephant greeting can also initiate reconciliation, deescalating friction after an altercation between two bonded individuals and reinforcing peace within a group. You may not have thought much about the importance of our own basic greeting rituals, but a simple hello or handshake is actually an opportunity to strengthen bonds, pay respect, or even calm a brewing feud.

For many reasons, modern society has made it easy for us to deprioritize rituals in our own lives, whether because we are too busy or because technology makes it easy to disengage and distance ourselves from our natural way of life. However you feel about social media, video games, TV, and other entertainment, they often leave us lacking direct in-person human contact.

There's a reason why solitary confinement is the worst kind of deprivation we can inflict on a social animal. Social animals evolved in the context of others. Visual and vocal communication, along with touch and proximity, are necessary for overall physical and mental health. Without socialization, a social animal withers and dies—humans included.

Our most natural state of social interaction is touch or physical proximity. Without it, we are missing something fundamental to our existence as social animals, including many physiological benefits. Lack of connection to a community can cause us to indulge in self-destructive behaviors that leave us feeling depressed and lonely. Thus, today, stress-related diseases are at an all-time high. Evidence points to a decline in ritual and real human connection and to an urgent need for social and psychological readjustment.

Social media and technology are a double-edged sword. While they offer opportunity, discovery, and the chance to forge new relationships, they also cause isolation, self-criticism, and alienation. All of these things can lead to potential mental-health crises. The magnitude of this situation was highlighted and made worse during the 2020 coronavirus pandemic, when vast numbers of people around the world had no choice but to isolate themselves and turn toward online social interaction.

After months of being quarantined in our homes, we were forced to recognize the impact of losing in-person social connections. Psychologists predict that the trauma people experienced through physical isolation will be transgenerational. The coronavirus quarantine revealed our strong need for real human contact.

At the same time, it revealed the vital role our daily rituals play in our everyday lives. They became the anchors that grounded us during tragedy and that connected us to one another even across oceans. Our frenzied, fast-moving world of places to go and people to see shrank to a short list. Without constant distraction and endless plans, we sang from our balconies together, had socially distanced conversations with neighbors over the fence, gardened, joined online baking groups and made sourdough bread, even saluting healthcare professionals with a chorus of sound every evening—all to participate in rituals of perseverance and hope.

Rituals are a lifeline during a crisis, but at all times we need fulfilling rituals to avoid feeling disconnected and alone. Even in this high-technology era, we remain inherently social animals, and what we seek is true connection in whatever form—with a stranger, with a neighbor, with colleagues, loved ones, and family. Returning to our roots, reconnecting with our wild side and the rituals of the wild can lead us down a path toward fulfillment, compassion, and well-being.

introduction

Reconnecting, with ourselves and others, through ritual, is a step in the right direction. Certain rituals have endured evolution, near-extinctions, and changing environments across the animal kingdom. By learning from these powerful ceremonies within the wild, we, too, can access our own innate ability to find healing, self-awareness, community, and ultimately deep connection to the world around us.

THIS BOOK WAS born out of a desire to show how and why ritual is critical in our everyday lives, using examples from across the animal world, including elephants, chimpanzees, orangutans, wolves, dogs, lions, zebras, whales, flamingos, fish, and even insects. Although social animals exhibit many rituals, this book focuses on ten important rituals that are essential to our well-being: greeting rituals, group rituals, courtship, gifting, spoken rituals, unspoken rituals, play, grieving and healing, renewal, and travel and migration.

Ritualized greetings, such as mouth-licking in wolves or a handshake in humans, are a form of information gathering, and they have evolved among social animals to strengthen bonds and build trust. Vocal rituals, such as battle cries before war or sporting events, generate a sense of unity and common purpose. These battle cries mirror the calls of other primates, like the roars of howler monkeys defending their turf at dawn and dusk, or the territorial roars of a pride of lions. Vocal rituals can also be a tool to release aggression and create boundaries.

Vocal rituals are used to unite a community, such as an elephant family's coordinated calls while departing from a waterhole. These coordinated vocal bouts are similar to our own highly coordinated duet songs or synchronized orchestral music led by a conductor. This coordinated group effort stimulates stress-releasing endorphins and a stronger sense of community in the musicians and vocalists.

Nonverbal rituals, such as smiling and laughing, have been around for more than five million years. These contagious behaviors have a positive impact on our mood, which is stimulated by the neurotransmitters dopamine and serotonin. These same mood-altering chemicals are also exhibited in our cousins the chimpanzees, as well as many other animals.

The simple act of gazing into another's eyes is a powerful ritual of courtship and other forms of bonding, both in the wild and in our daily lives. It builds intimacy within romantic relationships as well as the relationship between a parent and child.

The ritual of gifting offers many benefits, while play affords individuals the opportunity to experiment with their surroundings and come up with creative solutions critical to a species' survival. Examples include a lion cub practicing hunting skills by pretending a littermate is prey or a toddler building a fantastical world in a sandbox, which fosters coping skills for later life challenges.

Our rituals of grieving and healing from the loss of a loved one, including carrying or burying our dead, are also present in other species, such as dolphins, chimpanzees, and elephants. Renewal, in the form of welcoming a new season or spring cleaning, has some surprising origins with unexpected benefits. Finally, most would recognize the curative power of a pilgrimage or a simple change of scenery as a form of resetting one's perspective.

Each of these ten wild rituals are consistently positive and play a powerful role in the lives of social animals. Although the main advantages of some of these rituals may seem obvious, they also have many other subtle impacts that enrich our lives in important ways.

These ten rituals also have components that are missing within our current society—or have been altogether lost. What might seem like antiquated practices can actually enhance our physical and mental well-being. They are the key to successfully communicating, nurturing, and building stronger communities. Reincorporating the lost art of ritual will better equip us to discover new ways to reconnect to others, to ourselves, and to the natural world.

introduction

Many societies are deeply divided today, whether by race, class, age, income, religion, or gender. So it behooves us to do everything we can to understand and strengthen relationships within and between groups of people. We have the opportunity for renewal by reconnecting with important primal instincts and reacquainting ourselves with rituals of inclusivity.

As C. S. Lewis so eloquently wrote in *A Preface to Paradise Lost*, "When our participation in a rite becomes perfect, we think no more of ritual, but are engrossed by that about which the rite is performed; but afterward we recognize that ritual was the sole method by which this concentration could be achieved."

Some rituals are not important in and of themselves, but the outcomes are. They are responsible for helping us to build healthy lives and relationships. When we repeat big or small acts of kindness and inclusivity over and over again, they become part of who we are—and that is the purpose of rituals.

In fact, I must admit that with my deadline for this book looming, I almost canceled my plans to see family over the winter holidays. Then I reread the chapter on greeting rituals, which reminded me of how significant the act of acknowledging one another can be. An in-person reunion, no matter how brief, is much more important to our well-being than, say, a missed deadline. I immediately booked a flight home for the holidays to be with my family and felt so much better for it. As for my deadline, the schedule only slipped by *a few* days.

This book is an exploration of shared rituals that occur within both wild and human societies and offers a path to building and reinforcing rituals that add more meaning to our lives. Having an awareness of our habits and how they positively impact our well-being is important. Our good habits, no matter how small, can change our lives for the better. Reconnection with the rituals that have held our ancestors—and all living beings—together across time will bring us more peace and fulfillment.

1 · GREETING RITUALS

SPIT, SNOT, AND OTHER SOCIAL GREASE

"The light within me bows to
the same light within you."
—SANSKRIT NAMASTE GREETING

DURING THE MIDDLE OF A recent field season in Namibia, Big Momma, the matriarch of the African Queens, a prominent elephant family in the region, suddenly found herself separated from her group. Big Momma and her family had been enjoying spending time at the watering hole one particularly hot afternoon when a new family appeared in the clearing, and began to slowly edge them out. Big Momma's formidable phalanx of adult females marched toward the newcomers with confidence as they tried to drive them off. This is when something went horribly wrong, as a young matriarch of the new family charged at the group, scattering Big Momma and her family—a most surprising outcome. Showing signs of going into estrus, Big Momma caught the attention of a young male, which seized the opportunity in the chaos to chase her out of the clearing. This is how Big Momma ended up by herself, running through the brush in an endeavor to dodge her unwanted suitor.

It wasn't more than a half hour of separation, but it was an extremely tense situation and the family reunion for Big Momma's return seemed to be an enormous relief. It was somewhat chaotic—Big Momma running toward the family with dust and snot flying as younger females ran to join the adult females in greeting their leader with jubilant rumbles. They gathered around her, almost sandwiching her between them, and rumbled in long low-pitched tones replete with many trunk-to-mouth greetings, consolatory gestures indicating that they had no intention of letting such a separation happen again.

For female elephants, being separated, for any reason, is a risk, which is why their reunions can be such monumental occasions. It is also why bonded groups stay within vocal contact while foraging. Remaining in contact makes it easier to find one another in case a family needs to reconnect in order to help ward off a predator or any other threat to their group.

Big Momma's reunion made a strong impression on me. It showed just how vulnerable individual female elephants and their calves must be on their own. It also demonstrated how insecure a family might feel without its matriarch in its midst, knowing that she might be in trouble somewhere, in the distance. Regardless of how long they are separated, elephant family groups come back together with celebratory greetings. These greetings serve as a reaffirmation of the bonds these individuals have with one another.

GREETING RITUALS EVOLVED within groups of social animals for three purposes. One is to reinforce bonds between two close associates or a group of associates or to welcome a new friend. Another is to reduce tension and foster reconciliation. The third is to signal submission to a dominant individual, which promotes cooperative and peaceful coexistence within a society.

By performing a suite of often risky and intimate acts, one can effectively communicate one's intentions to engage in further affiliative exchanges. Completing actions that require mutual trust within a greeting ritual can also offer a way to assess an individual's potential as an ally.

For example, both male and female hyenas present their erect genitalia during a greeting ceremony, a very vulnerable position, indeed. Studies of hyena social structure indicate that reinforcing trust in existing relationships, through this greeting ritual, facilitates coalition formation within hyena groups. This helps them to be more cohesive, such as when they engage in clan wars or have to push lions off their kill.

More like our handshake greeting, the elephantine ceremony between Knob Nose and Donut, and Big Momma and her family, reminds me of how easily we tend to dismiss the importance of a proper greeting in our busy, technologically driven lives. Compounding the issue, in situations where there is frequent interaction, we might mistakenly assume that greetings are not necessary at all, thus forgoing thousands of years of human ritual.

Today, in many situations, daily greetings on the street or subway—or in some situations, even with a neighbor—are now awkward, such that we scarcely attempt or even actively avoid them. In allowing this "greeting fatigue" to be our new norm, we become isolated from our communities.

The act of saying hello, giving someone a smile in passing, making eye contact, or offering to shake hands may feel insignificant. These simple acts, however, have been shown to be vital to our well-being. Greeting rituals generate an air of positivity and connectivity within our communities. They improve existing relationships and help forge new ones.

I often think about how in more traditional societies that don't have much technology (let alone electricity), direct interactions with people are constant and greeting rituals remain common. I spent some years in the early 1990s, and again twenty years later, working in Zambezi (formally Caprivi), a remote region in Namibia, with a number of very traditional societies. Greetings in these societies are paramount to one's social existence. I quickly learned that the greeting ritual included a number of critical steps, depending on the relationship of the greeting parties as well as their position within society.

Over the course of my many social interactions in Zambezi, I was exposed to several levels of greetings. In my work protecting farmers from crop-raiding elephants, I became very close to the women farmers. This led me to get involved in community development and female empowerment issues, as well as issues surrounding land disputes and the HIV crisis in the region.

Within these many contexts of working with people, I learned that the most casual greeting between associates in Zambezi includes a style of hand-over-hand clapping with a slight bend of the knee. The greeting also includes a spoken salutation, depending on the time of day and the language of the greeting party. The clapping and knee-bending can be repeated several times during the salutation, which confers additional respect depending on the relationship. A greeting for close associates might also incorporate a three-part handshake, involving a standard shake, then a clutch, and then a handshake again, followed by another hand-over-hand clapping sequence with a slight bend of the knees.

The highest-level greeting is reserved for either the mayor (*InDuna*) or the chief. This includes a full crouch (as low as one can possibly muster) during a bout of hand-over-hand clapping. Once this portion is completed, you take a few steps forward and start the ritual again. And if you happen to be a woman, there is also a tripping hazard, as women must wear a long skirt to appear before a chief—the crouching and clapping and then standing, stepping forward, crouching, and clapping again requires a dexterity little appreciated by non-skirt-wearers.

As you might imagine, this greeting ritual can be a somewhat time-consuming endeavor, particularly if there is a whole group appearing before the chief. Each specific act performed in this very detailed sequence serves to convey one's recognition of the chief's status. The use of this cultural greeting honors their social traditions.

Late one night, while my colleague and I were attempting to track down a group of troublesome male elephants, known to be raiding crops along the river, I received the most honored Zambezi greeting of all. We had stopped at a local *khuka* shop (a local bar) to find one of the game guards who might have knowledge of the offending elephants' whereabouts.

As soon as I got out of the truck, I was surrounded by a group of very old women who seemed amused by my presence. They laughed at me through sparse, tobacco-stained teeth. Some were friendly laughs, some seemingly half-mocking—as a *makua*, or white person, wasn't always welcome in such settings.

Several of the women wanted to touch me, and one grabbed at my hand, while my colleague tried to wave them away. When I hesitated to leave, he dismissed them as drunken troublemakers. It was immediately apparent how these women had spent the day—or most likely several days. Since there is no refrigeration in the region, when the local traditional batch of homemade fermented grain (beer) is ready, everyone becomes drunk for a four-day period.

I didn't want to deter the enthusiastic desire of these elders to engage with me, so I allowed one of them to open my palm. She held my hand and spread it open further as if she were going to do a palm-reading. She laughed and mumbled incoherently as the others closed in, forming a circle around us. She struck a more solemn tone, as if she were about to perform some kind of rite.

I tried not to look intimidated when suddenly the woman started spitting into my palm. I couldn't help feeling a little concerned that this interaction wasn't entirely good-natured. I resisted the urge to flinch as the woman continued to spray spittle into my opened palm with great ceremony, while the others huddled around us and looked on unabashedly. The others didn't participate in the spittle contribution, thank goodness, but they provided no clue as to what was going on. Hoping the mystery was going to be solved imminently, I continued to provide my saliva-soaked palm for these women's entertainment.

At that point, the game guard appeared and smiled. He explained that this woman was bestowing a great honor upon me by calling up her ancestors to greet me. The spray of spittle represented splashing water onto the embers of a fire—the local method for summoning ancestors. I smiled at her warmly, relieved that there was a positive end to this uncertain exchange.

Being unfamiliar with another culture's greeting ritual can easily lead to misunderstanding. All social animals have some form of greeting ritual for a reason. *It is a sign of recognition, goodwill, and welcome.*

Often a greeting expresses different levels of respect depending upon the relationship between individuals. The European "air kiss" on both cheeks signifies a special relationship or occasion. Many are probably familiar with the nose greeting by the Inuits called the *kunik* (what was once referred to as the "Eskimo kiss"). This greeting ritual involves pressing the nose and upper lip against a family member or loved one's cheek or forehead and breathing in their scent. A very similar greeting ritual exists in Polynesian and Hawaiian cultures, where two people press noses, or noses and foreheads, while inhaling together the *ha*—the breath of life—and *mana*, or the spiritual power between two people.

If an Inuit friend in Greenland leaned in to give you a nose rub, or a Maori friend in New Zealand, or someone of Hawaiian descent in Hawaii, you would very likely feel honored by this reverent cultural welcome. Outside of these cultural contexts, if someone pulled your nose and forehead toward theirs and breathed in deeply—say on the streets of New York City—this gesture might seem offensive.

An acceptance of different cultural practices is key to creating new relationships around the world. When communicating with another society, learning the appropriate greeting ritual is the most important—and often the safest—place to start.

A submissive greeting allows one to acknowledge their place in the social pecking order. This action helps to minimize potential conflict and reduce stress in a group. All cultures have special greetings that show deference for elevated status. For example, if I were to go to Buckingham Palace to meet the queen, I had better brush up on the eight-step greeting ritual for British royalty. The first step is to rise when the queen enters the room and remain standing unless directed to sit, or until she sits. Next is a brief bow or curtsy that is required of a British citizen. Third is to use the salutation "Your Majesty." To show respect when dining in the presence of the queen, one must wait for the queen to eat first and then eat in silence.

For male elephants, the purposeful act of placing a trunk in a dominant male's mouth is akin to the prime minister kissing the queen's ring, or kissing the ring of the bishop or of the *capo di tutti capi*, a mob boss—even to the point of lining up and performing the act in single file, usually in order of dominance.

There are many more human cultural greetings, as well as an infinite number of nonhuman animal greetings. Take, for example, the gorilla or chimpanzee hug, the bonobo kiss, and the zebra nip. I like how other great apes hug because it is so similar to our own— two bonded gorillas or chimps relaxing into each other's arms. The bonobo kiss on the lips is an even closer parallel. I also like the zebra nip for how playful and spirited it is. It can also be tense—with two young males ready to spring into action.

Mid-July at Mushara waterhole is a very dramatic time. Zebras arrive en masse, as water availability south of this waterhole diminishes, and zebras depend on Mushara to drink.

In the mid- to late afternoon, hundreds of zebras spill into the clearing. Family by family, the dominant females and their offspring follow in a line through the sandy track, head to rear. Their heads bob as they shuffle along, while the stallions of each harem peel off to participate in greeting ceremonies. Young males within smaller bachelor groups burst into the clearing behind these organized families. Each is eager to greet and play and test out their prowess with other young males.

The lead stallion of each harem will seek out other harem stallions and form little greeting parties. They call out in vocalizations that are quite noticeably distinct from one another. These vocalizations are unexpectedly high-pitched emanating from such fantastical-looking creatures. After announcing themselves, the harem stallions nuzzle, wrap necks, nip, and sniff one another around the head and sometimes the genitals. Then they exhibit an exaggerated chewing behavior with their teeth bared and the corners of their lips drawn up as if smiling or laughing with necks extended and ears directed forward. It really looks like they are enjoying a hilarious joke together. After these exchanges, the dominant stallions gather around the defecation of one of the other males and inspect it before all of them defecate together, forming a communal dung pile.

Upon arrival at the waterhole, the young stallions from various bachelor groups start to nip at one another's necks in greeting, inviting one another to play. These nips are playful and often accompanied with head-bobbing and a high-pitched call. The tense threat of a kick following a nip is full of kinetic energy that could result in a variety of outcomes—including giving chase and trying to knock the other off balance. This dynamic can escalate into more aggressive play, including raising up on hind legs and kicking at each other with front legs while attempting to bite the other's neck. All these scenarios start with a simple nip.

The greeting ritual between male zebras is all about defusing tension and building trust. For males in many species, greeting and play are sometimes intertwined. The invitation to play is either incorporated into or quickly follows a greeting. In play, male zebras can further demonstrate that they have no intention to harm by engaging in playful behaviors, like nipping without inflicting injury.

Male black rhinos will sometimes greet by touching horns. This initial contact might escalate into a gentle spar back and forth, the slow crossing of horns on one side and then the other—similar to two swordsman clacking swords back and forth. This usually involves some wiggling or flattening of the ears, the flattening of ears being a signal of either curiosity or submission.

As the two black rhinos back away from each other, the air is cleared by this initial contact and they proceed to take a drink at a safe distance, while keeping a careful eye on each other. Or, the mood can sometimes turn even tenser if one perceives that the other somehow botched the salutation and proceeds to vocalize in a loud bellow.

This often instigates the beginning of a long night of bellowing, heavy breathing, and shuffling of sand to define boundaries, and could escalate into a bloody battle in the middle of the waterhole—which has on the rare occasion resulted in mortal wounds. Black rhinos are known to be solitary after all. This little bit of horn contact appears to be just about as much as they are willing to entertain. But that one touch of the horns serves its purpose of diffusing tensions—most of the time.

A low-ranking male black rhino at Mushara that we affectionately nicknamed Scratchy because of the deep scratches he incurred from a run-in with lions avoids these potential escalations by coming to the waterhole during the day. Most black rhinos tend to drink at night. For the low-ranking Scratchy, this is apparently a safer strategy than risking this dangerous greeting ritual.

OUTSIDE OF SCRATCHY changing his schedule to avoid a potential rhino scuffle, animal societies do not appear to exhibit "greeting fatigue" like we often do. An obvious example is a dog's greeting ritual. There isn't a day when my dog, Frodo, would skip the opportunity to greet me coming home from work. Usually his daily greeting entails an elaborate display of exuberance, involving prancing and bowing, running in a circle with ears perked and eyes wide, combined with spirited tail-wagging. Jumping, licking, and groin-sniffing are suppressed as best as he can, as he knows these behaviors are discouraged. Upon particularly enthusiastic greetings, there is even the presentation of a gift—a favorite squeaky toy—or the giant stuffed Clifford that he picked out from the "free" pile at a garage sale.

Having evolved from wolves, domesticated dogs have adapted their greeting rituals to suit their human companions. For the most part, their displays are similar to wolves—especially face- and mouth-licking, whining, and groin-sniffing. Frodo's greeting rituals have forced me to consider what we, as humans, might be missing by being so comparatively blasé about our greetings.

I first noticed a decline in human greeting rituals in the early 2000s, while living in high-tech Silicon Valley. My husband and I were assigned to graduate student housing on the Stanford campus at the height of the dot-com boom. Despite being in this small collegiate community, people were living in their own little isolated worlds. No one looked others in the eye or said hello in the elevator.

One might expect such behavior in a crowded elevator with anonymous strangers in a New York City skyscraper. It seemed strange that such active avoidance was exhibited in our twelve-story apartment building filled with graduate and medical students, who were all attending the same university in a small town. Once, I forced myself to have an exchange with a neighbor while riding the elevator, and we formed a fast friendship. This friendship lasted the length of our tenure in that building and made the cramped living conditions far more bearable. Years later, I realized just how much all of us could have improved our community and emotional well-being by engaging in simple greeting rituals. At the very least, eye contact and a nod of acknowledgment would have gone a long way toward fostering a sense of community.

Greetings are the first moment of connection across all societies, human and nonhuman. They strengthen bonds and build trust while also providing real-time data about another individual's hormonal or psychological state. For example, the trunk-to-mouth greeting of elephants evolved from a form of information gathering. Placing a trunk in another's mouth is a way of finding out what the other individual has eaten. Since elephants are not born with the knowledge of which plants are safe to eat, they need to learn this information from others. And it makes sense that such an important behavior would also have become ritualized into a gesture of greeting. The same is true for the wolf and mouth-licking. What had once been a form of information-gathering was ritualized into a greeting that stands apart from the act of learning what another might have eaten. These greetings also provide an opportunity for an exchange of hormonal information about the other's emotional and physical well-being.

Humans exchange hormonal information in a handshake. The handshake has been around for thousands of years. And its lasting power may be in part because it serves the same purpose as other tactile greetings—to assess the hormonal status of the other person through bodily scent. In one recent study, subjects were inclined to sniff their hand, or to touch their face with their hand, immediately after a handshake. This action was exhibited more with same-sex than with opposite-sex handshakes, indicating that a handshake might be more related to an assessment of dominance than to finding a possible mate.

The human handshake evolved for other reasons as well. It was first used in Greece in the fifth century BCE. Presenting an opened hand was a symbol of peace, since it showed neither person was carrying a weapon. In the Roman era, the gesture evolved into a forearm grab to check that neither was hiding a knife up their sleeve. The shaking gesture may have developed among knights in medieval Europe in an attempt to shake loose hidden weapons.

Meanwhile, in the United States, eighteenth-century Quakers adopted the handshake greeting—instead of the curtsy and bow, or any gesture that recognized hierarchy, dominance, or status—in their effort to emphasize that everyone was considered equal.

The modern French kiss on the cheek, *la bise*, has a similarly intriguing historical evolution. Used in religious ceremonies in early Christianity, it became a symbol of trustworthiness in the Middle Ages whenever people entered a contract. Then, during the plague, the practice was dropped, and it did not return until after the French Revolution, four hundred years later. This kiss has since been temporarily avoided during modern pandemics, such as the 2009 swine flu and the COVID-19 outbreak in 2020.

Greeting rituals change over time. Some traditions get lost or abandoned, for one reason or another, but *what never fades is the importance of greeting rituals themselves.*

The simple act of greeting has a powerful impact on those involved, equally so when it's a stranger. Greeting a stranger with eye contact, a smile, and a hello feels rewarding, especially if we get a smile in return. In fact, scientists have found that when a stranger returns a smile, this fills us with a sense of positivity. Also, apparently, we are happier on days when we have more interactions with acquaintances because, oddly enough, we feel more connected when we talk to strangers than people we know.

This kind of engagement forces us to explain and speak in more depth to convey our thoughts. This is healthy because we are required to think more about our emotions than when talking to someone who is familiar with us. This fosters a deeper understanding of ourselves and our experiences—similar to talking to a therapist or within group therapy for the first time.

Additionally, a positive encounter with a stranger gives us a sense of control because we can decide how much information to reveal. In fact, sometimes it is easier to open up to a complete stranger.

Resonating with someone we have just met gives us a sense of connection and meaning because it validates our emotions.

This positive exchange with strangers is thought to be an adaptive behavior that evolved over the course of human history, since mating outside our own gene pool is advantageous. This means that our social skills, which facilitate our interactions with people outside our immediate circle, are actually survival skills.

These benefits are not confined to strangers, of course. Studies show that people who engage within community groups or organized social opportunities have better emotional and physical health and live longer than people who do not engage in these activities. In fact, a growing number of scientific studies highlight the importance of physical and proximal social experiences. All of these benefits start with a simple hello. Without greeting rituals as a tool to grease the social wheel, we might never get to these more intimate opportunities to engage.

Watching Knob Nose and Donut—a mother-and-daughter elephant duo—reunite, and Big Momma rejoining her family with such an intense greeting ritual is a reminder of just how important reunions are to reinforcing connections with friends, family, and others. It might seem simple or obvious, but saying hello is a life-saving exchange.

Fostering the ritual of greeting is good for your health and opens your mind and heart. We are social creatures after all, and if we neglect or avoid social interaction, we do ourselves a disservice. Greetings can be simple or elaborate, quick or time-consuming—but they should always be mindful. Look someone in the eye and smile, shake hands and hug, offer an elbow or fist bump, bow, kiss them on either cheek, touch foreheads, or even—if it's culturally appropriate and custom or circumstance calls for it—give them an enthusiastic spray of spittle in the palm of the hand. We will be stronger for it.

2 • GROUP RITUALS

THE POWER OF THE COLLECTIVE

> "Alone we can do so little.
> Together we can do so much."
> —HELEN KELLER

EMPTIED THE AIR OUT OF my buoyancy control device and lay on my back looking up at the surface of the water twenty feet above. Light danced across my wetsuit and over the rippled white sand at the bottom of the lagoon.

Next to me, the seafloor dropped abruptly, forming a wall that fell about 100 feet to a second sandy bottom. I breathed in and held my breath so no bubbles from my regulator would interrupt my commanding view of the ocean, which pulled like a magnetic force, away from the comfort of White Sand Beach off the private Guana Island in the British Virgin Islands and toward the deep blue depth beyond the shelf.

I clasped my hands behind my head and looked up through the water at the wavering sun. A large dark shape caught my eye as a spotted eagle ray glided overhead. With a few flaps of its wings and its long tail trailing behind, it disappeared into the depths a hundred yards away.

My eye was then drawn to the anchor line going up to the surface. I wondered why my husband, Tim, was taking so long to drop off the pieces of elkhorn coral on the boat. We had collected pieces near Guana Point that had broken off in the last hurricane. The large branches of coral would most likely die, rolling around in the sand as they were, so we planned to glue them back to the shallow finger reefs in White Bay. We had started a reef restoration project there with some colleagues. We had been underwater for hours and I was starting to get chilly.

A black form amassed like a storm cloud roiling beyond the shelf. In an instant, I found myself engulfed in a dark wall of energy. The ocean became so thick with the silver glisten of thousands of anchovies I couldn't see the surface.

The sudden blackout and panic of fish just inches from my nose triggered claustrophobia. However, as much as a part of me wanted to break through the barricade to reach the light of day, I kept still as the dense school of fish passed silently overhead.

The water was now alive with pelicans diving through the living tangle of fish. Shafts of bubbles opened in the solid layer, letting pinpricks of light through the darkness. The tiny fish packed themselves into a denser and denser mass, whirls of current swirling out as they darted in one direction and then another. The storm of panicked fish persisted for several minutes before it started to get lighter again.

Just beyond the wall of fish, I saw the cause of the commotion—a huge school of giant tarpon cruised smoothly in close formation. Each of these menacing-looking, torpedo-shaped fish stretched almost the length of my body. As they corralled the anchovies above me, their silver-blue bodies flashed in the light before exploding into the living mass of food with their mouths opened wide.

When the tarpon group hunt finally subsided and the sparkling surface of the water was visible again, I breathed larger breaths. The trickle of air morphed into silvery blobs that merged and billowed up in a thin trail. I took one deep breath and exhaled long and hard. This time the air bubbles ballooned to the surface like giant bulbous mushrooms.

group rituals

The ballooning bubbles from my exhalations reminded me of how humpback whales hunt in Alaska, using a technique called "bubble-net feeding." A group of anywhere from several to sixty humpbacks position themselves in a circle underwater blowing a curtain of air bubbles to concentrate a small group of salmon or herring or krill for a much-easier meal. When the net is in place—sometimes extending up to a hundred feet—one whale will vocalize to announce that the hunt is on. Upon hearing this call, all the whales rise to the surface with mouths wide open, catching as many fish as they can.

Dolphins do something similar. One dolphin swims in a circle in the shallows, beating its tail against the muddy bottom to cause fish to get trapped inside the ring of mud. As the dolphin makes a tighter and tighter circle, causing the fish to jump out of the water to escape, the rest of the pod waits with mouths opened, ready to catch the fish.

One time, a group of friends and I responded to a snorkeling ad to see a famous wildlife spectacle: sailfish group hunts of anchovy schools off the coast of Mexico. We'd all seen stunning footage of predators like whales, sharks, dolphins, tuna, and sea lions following sardine migrations on documentaries such as *The Blue Planet,* and for one friend, the one who convinced us to go on this crazy adventure, it was an important box to tick on his extreme-nature checklist.

Little did we know that the tour operation we signed up for consisted of an old wooden skiff and a captain and scout who only spoke broken English; they divulged few details of our anchovy safari ahead of time. About a mile offshore, we and a group of seasick Russians entered the most gnarly oceanic water the Atlantic could pitch up.

It had been even rougher on our first attempt the day before, so rough that the captain decided to cancel after we had gotten out there and the swells were too big to navigate safely. So, it was with a little trepidation that we got out to the boat the second day, knowing what rough seas might lie ahead of us. Fortunately, it was slightly better this second time around.

When the captain spotted roiling patches of water with seagulls hovering above, he'd bring the boat closer to determine whether the feeding frenzy was caused by tuna or sailfish—the tuna hunts were much faster moving, apparently. When the conditions looked right for sailfish, the scout put on his mask and jumped in the water to have a look around, while we all tried as best as we could to hold our stomachs down.

Watching the scout swim into mysterious waters, unsure of what could also be lurking below in search of a meal, I suddenly couldn't help wondering what a great white feeding frenzy would look like. Yet when the scout gave the thumbs up, we all donned our masks and fins and flopped over the edge of the boat into the infinite blue liquid. We were completely entrusting ourselves to the captain and scout—and the sea, with its tangle of giant predatory fish sporting formidable swords.

In a matter of seconds overboard, we saw what we had come a great distance to see. Somehow the psychological consolation of safety in numbers, as well as of the clunky old wooden boat now some distance away, soothed my nerves enough to allow me to enjoy this otherworldly phenomenon. There they were—a school of magnificent sailfish corralling a bait ball of anchovies in a spiral with their dorsal fins positioned like sails, bowed out in the middle.

Separated from sardines by tens of millions of years of evolution, anchovies don't migrate nearly as far, and they group together in huge schools to engage in coordinated defenses to minimize the chances of getting eaten. Predators like sailfish counter this strategy by attempting to isolate a group of fish away from the larger school.

Sailfish look very much like blue marlin, whose dorsal fin is less pronounced. While the sailfish swam in a circle, their bodies and fins flashed a brilliant blue on and off, most likely to confuse the school of anchovies. They used their pectoral fins to steer in a tighter circle, enveloping the school of fish caught up in a mini-twister—until one of the swords burst through the twister.

group rituals

The sword made short work of the diminishing school, slashing through the dense column of fish, leaving scales floating where fish had been swimming moments before. When the fish from the bait ball tried to escape by going into deeper water, some of the sailfish went down and used their bills to change the trajectory of the fish back up into the shallower water. There, they sandwiched the column of fish against the water's surface.

This coordinated sailfish hunt was an experience of nature I will never forget. However, many bait balls later across miles of open ocean, we were happy to end this wild west sea adventure and return to land.

OVER THE DECADES that I have worked in Etosha National Park, I have seen many coordinated hunts within groups of carnivores, such as prides of lions, hyena clans, and cheetah families. I have seen coordinated mobbing, where a group of wildebeest chased away an exhausted cheetah from her resting site under a tree. The shared understanding among the wildebeest is that a predator should always be displaced, and it's much less risky to do so within a coordinated group.

My husband, Tim, and I once listened to a high-speed chase in the middle of the night, as a hyena clan hunted in a dense riverine forest in the Zambezi region of Namibia. Each member of the hunting party used demonic vocal cues to coordinate their position with the other hyenas while running.

Coordinated hunting in desert lions in the coastal zone of Namibia is tactically equivalent to soccer players taking their positions on the field. Carnivore scientist Flip Stander documented that each female in the pride has a specific position in the hunt that she always assumes. Each hunter has expertise similar to each soccer player. For instance, on a soccer field, the wing runs the ball down to the striker, who takes the shot. On the lion's playing field, the wing does the high-speed running to steer the prey toward another lion who plays the striker and makes the kill.

There are many other examples of cooperative and coordinated hunting, such as a wolf pack taking down a moose in Yellowstone or a group of chimpanzees hunting the colobus monkey in equatorial Africa. There are even cases of cooperative hunts between different species, such as the African honeyguide bird and the honey badger. The honeyguide leads the badger to a beehive, and together they share the spoils. This same cooperation exists between the honeyguide and different groups of African people like the Yao in Mozambique.

Cooperative-hunting behavior in forest chimps is thought to provide clues to the evolution of our own cooperative-hunting strategies. In fact, primatologists believe that group rituals may have evolved in human societies from the need for a group of men to cooperate on a hunt.

During early hominid evolution, cooperation within a large hunting party was necessary in order to take down enormous animals like mammoths and mastodons using only spears and ingenuity. These hunts required a tremendous amount of organization and combined effort. Coordinated hunting was a matter of survival at first, but over time, it may have led to successful cooperation in other areas of society.

Despite many traditional societies being displaced by economic and geopolitical events over the last century, subsistence group hunting still exists in small, indigenous hunter-gatherer societies. The San people of southern Africa; the Inuit of the Arctic and subarctic regions of Siberia, North America, and Greenland; as well as the Mayan peasant-hunters across the Yucatán Peninsula in Mexico—all have continued to hunt and share their tactics with younger generations.

Often traditional hunts are prefaced or followed by a spiritual ritual acknowledging the hunter's and the animal's special place in the world. The traditional Inuit believed that animals were superior to them but that the animals allowed themselves to be hunted, and as such, the Inuit performed rituals and songs to thank the animal's spirit.

The Loojil Ts'oon, or Carbine Ceremony, of the Mayan peasant-hunters is still performed today prior to a hunt. In this very involved group ritual, participants ask for divine permission to hunt an animal and acknowledge the animal that is about to be hunted. They prepare, cleanse, and deliver a soup to the spirit, *Sip*. Then the jaw of a deer is cleaned and carried to a particular place in the hills that represents the lord of the animals so that they may give these animals new life.

This group ritual serves to legitimize the killing of the animal as sustenance as well as to maintain an equilibrium between the hunters and the natural world. The group is conscious of the size of the animal population being hunted and of the need to use the resource sustainably so that it continues to provide sustenance to their communities.

Group rituals within both human and nonhuman animal societies are used for many different purposes other than hunting. They are used to define territories and boundaries, prepare for battle, and communicate over long distances. They are used to attract mates, to trigger courtship, to galvanize a group for a cause, and to engender trust and group identity. All of these activities require a level of coordination, cooperation, and often some level of synchrony that are obtained through ritual.

There is evidence of group rituals throughout human history, some immortalized through cave paintings and archaeological sites. Often these rituals are associated with religious rites, hunting a large animal, performing a sacrifice, or celebrating a change of seasons, a harvest, a rite of passage, a union, or a loss. As human populations grew, psychologists have proposed that group ritual also may have evolved as a mechanism to provide identity to individual groups that were increasingly composed of nonkin, and this may have led to the origin of cultures.

The evolution of group living was so beneficial for primates that it was thought to have resulted in the evolution of a larger brain, cultural complexity, and language in humans. In fact, in the last few million years, our ancestors' brains tripled in size. The relationship between brain size, group size, innovation, social learning, and culture all contributed to the rapid expansion of the human brain. In theory, brains had to expand to store and manage greater amounts of information, and more information meant a richer, more complex culture with a more sophisticated means of communicating. The advent of language, then, catapulted humans on a different evolutionary path from our great ape cousins.

Individuals could also gain the loyalty of others by participating in a common ritual, like an initiation ceremony in which participants pledge their commitment to the group. This display can show dedication to the same values of the community and even reduce the possibility of aggression within the group.

Living in a collective offers other enormous benefits to all social animals. Coordinated caring of offspring and leveraging group knowledge maximizes the likelihood of survival while also passing down important life lessons. Sometimes we take the transfer of knowledge for granted because it is so ubiquitous in humans.

Other social animals offer striking examples of upholding this incredible ritual. During a severe drought in Tarangire National Park in Tanzania in 1993, older elephant matriarchs that had lived through a previous drought thirty-five years earlier were able to show their families an ancestral place to find water. These families had less calf mortality than those with younger matriarchs that had not previously experienced drought. A mother chimpanzee teaches her offspring how to collect termites by passing them a specially made termite-fishing probe made from the flexible branch of an herb, which they fashion into a brush-tip to maximize collection. Inuit grandparents play an important role in teaching their grandchildren and future generations the survival skills of hunting seals, narwhals, and polar bears. In this way, collective knowledge allows us to survive in times of scarcity by learning from those who have the most experience.

Through generational and collective knowledge, scientists can come together to pool their expertise to discover global solutions for viruses and diseases. With regard to morality and peaceful coexistence, North American indigenous groups pass down the Seven Grandfather Teachings to facilitate harmonious group living. These teachings include humility, bravery, honesty, wisdom, truth, respect, and love.

Taking part in group rituals can produce specific chemicals such as endorphins that affect our brains, stimulating feel-good sensations like happiness and even euphoria, or what runners refer to as a "runner's high." A runner's high is a complicated cocktail of neurotransmitters, including cannabinoids (also found in marijuana) and endorphins. Endorphins also reduce the sensation of physical pain. One study of male rowers from Oxford University showed that individuals could tolerate twice as much pain while working as a team than alone. Research also shows that group laughter can increase pain thresholds.

On the other hand, shared rituals that foster intense fear and anxiety, like a fraternity hazing or army boot camp, elicit dysphoria, which is the opposite of euphoria. Dysphoria causes each member of the group to experience what psychologists call *identity fusion*. The bond that forms during such an extreme shared experience generates a sense of oneness that is so strong it can lead to self-sacrifice, including a willingness to fight and die for the group.

Other examples of dysphoric rituals are coming-of-age or rite-of-passage rituals, often designed for young males as they enter adulthood. Take, for example, the male group ritual of land diving on Pentecost Island in Vanuatu, called *naghol*—the early inspiration for bungee jumping. During the dry season, as part of a harvest ritual, a group of men jump off a tower sixty-five to a hundred feet high with only a vine attached to each ankle.

Ten to twenty men will jump, attaining speeds of around forty-five miles per hour before reaching the end of their vine. Only their hair is supposed to touch the ground. The ritual was originally designed to thank the gods for a successful yam harvest and to bring good luck for the following year's harvest. Later, it also served as a rite of passage for young boys to demonstrate their bravery.

The *naghol* ritual includes a number of steps. Before the event, the men build a platform and prepare vines. The morning of the event, divers partake in a ritual wash, cover themselves in coconut oil, decorate their bodies, and wear boar tusks around their necks. The whole community dresses in ceremonial garb and watches from the ground, while many sing in order to encourage the divers. A lot of coordination and cooperation within the community is necessary for this ritual to be performed safely.

In a study of fire walkers in a Mauritian Hindu community, the fire walkers reported greater happiness after having endured the ritualistic suffering, which would be expected. Even more interesting was that those who witnessed the ritual were more exhausted than the fire walkers themselves. This exemplifies how the level of involvement in a group ritual shapes one's experience, both physically and psychologically. *Witnessing a ritual can be equally as powerful as performing it.*

I have witnessed a similar dynamic numerous times among elephants at my study site at Mushara waterhole in Namibia. After a male in musth mates with an estrus female, the entire elephant family, which has been charged up for this primal occasion, will engage in postcopulatory screaming and rumbling. This important bonding ritual is an otherworldly display of vocal excitement, and appears to have a powerful physiological effect on all involved.

One such occasion transpired on the night of a new moon. A mob of young males arrived chasing a young female from one of my study families. She was exhausted and distraught, with her family following

group rituals

behind. With the arrival of the dominant musth bull, Smokey, the gang of young male hoodlums took off.

Smokey was much more gentlemanly in his pursuit of the estrus female. Then, after Smokey and the young female copulated in the middle of the watering hole, the family remained abuzz with screams, roars, and rumblings that went on long into the night.

For elephants, these jubilant calls from the collective might facilitate ovulation in a similar way that repeated cooing does in ring doves, which you will learn about in the next chapter on courtship. After seeing this behavior a number of times, it has made me wonder if the low success rate of births of African elephants in captivity might be due to the sterile social context and lack of group chorusing during the union of elephantine sperm and egg.

This ritual, along with countless others, shows the natural instinct social animals have to participate in the group, whether it be for their own benefit or that of their herd or pack. A recent study demonstrated an incredible feat of cooperation among captive wolves. The wolves were trained to pull on a rope to get a treat, but the catch was that each wolf had to pull on a separate rope at the same time in order for the treat to be dispensed. Yet the wolves were able to successfully accomplish this fairly complicated joint action to retrieve their prized snack. Elephants have also succeeded in a similar test, in which they needed to cooperate to achieve the goal and receive a treat.

The domestic dog, meanwhile, has failed this challenge miserably. Most likely this reflects domestication. Dogs have evolved for so long with humans, and been separated for so long from wolves, that they seem to have lost their ancestors' skill of pack-wide coordination.

I have also seen many examples of elephants cooperating to rescue another individual. Mothers will coordinate a calf rescue, but with varying degrees of success depending on age and the skill and experience of the group. On one occasion, the matriarch, Big Momma, and her daughter, Nandi, synchronized their movements by bending down onto their knees at the same time in order to wrap both their trunks around a large calf. In this position, they easily pulled the calf out of trouble. If elephants are not bonded or don't have much experience coordinating, a calf rescue doesn't go nearly as smoothly. In one harrowing incident, a calf became stuck in the outflow of the pan, and the mother had no idea how to get it out. The rest of the family panicked at the sounds of the calf bellowing for help. The mother raced back and forth, trying to pull on her calf's trunk and steer it in the direction she thought best, but each time she failed.

Watching, we became particularly nervous, knowing that lions were in the vicinity. Fortunately, after much anxiety on the part of both elephants and researchers, the mother finally prevailed in rescuing her calf. This almost-botched rescue made me appreciate the ability of Big Momma and Nandi to coordinate and synchronize their actions. They made success seem so easy.

GROUP RITUALS OFTEN combine elements, like repeated vocalizations and synchronized, repeated body movements. Human group rituals, particularly religious ceremonies or community celebrations, often center around dancing, singing, chanting, playing music, and reciting a prayer or poetry. Sometimes this involves synchronous hand motions and strenuous physical exercise. There's a reason for this. Studies have shown that group rituals involving synchronous movement, or synchronous singing, inspire a feeling of connection as well as increased trust of others within the group. The repeated physical movements also trigger the release of oxytocin, the hormone that produces positive feelings of love, bonding, and well-being.

Nonhuman animals also engage in many types of synchronous group behavior. In the wild, elephants coordinate the timing of leaving waterholes using synchronized vocalizations. A dominant individual initiates departure by stepping away from the water and emitting

group rituals

a long, low-pitched rumble. In a coordinated, synchronized response, other individuals respond in a turn-taking manner, vocalizing one right after the other, with a small overlapping segment at the end of each call. Not only would this longer signal travel farther in the environment to alert others to their imminent departure, this repeated call helps to gather the group and align their collective intention and action. This vocal exchange always reminds me of a symphony conductor signaling the next group of instruments during an orchestra concert. The behavior is *that* coordinated.

Soldiers marching or marching bands performing for school sports or holiday parades display an impressive amount of synchrony. An awe-inspiring example of synchrony occurred during the opening ceremony of the 2008 Summer Olympics in Beijing, when 2,008 drummers emerged playing bronze Fou drums. Between the euphoria of the drummers, and the awe and wonder of the audience, the synchronized music and movement did the job of galvanizing everyone to feel part of this worldwide group celebration of international cooperation and sports.

On April 22, 2017, my husband and I participated in the inaugural March for Science in Washington, DC. It took a little while for a critical mass of people to form for the march, causing my husband to become skeptical of our effort. The rain and cold were additional deterrents that seemed to dampen enthusiasm. However, within half an hour, our peaceful crowd of like-minded folks was walking elbow to elbow down the middle of Constitution Avenue. As we passed the National Museum of Natural History, we chanted, "Science, not hate, makes America great. Science, not hate, makes America great."

Choreographed clapping accompanied our chant and deepened the experience. It reminded me of my high school cheerleading days, when I stood in the bleachers with my fellow cheerleaders, cheering on our basketball or football team. We would chant the lyrics to the

Queen song "We Will Rock You," while stamping and clapping the beat in between refrains.

The thrill of intimidating the opposing team with the power of our song and pounding feet and hands was palpable, as was the solidarity this generated with and for our team. The comradery intensified every time our team scored—and became ecstatic when we won and left the bleachers victorious and chanting another Queen favorite, "We Are the Champions."

As the march continued, I could see my husband's doubts about its importance fade. I couldn't help feeling that we were taking part in something much bigger than ourselves. Holding signs that read "There is no Planet B," Tim and I were surrounded by the vibrant, powerful energy being generated by the sheer number of people. We weren't alone; we had all gathered as a group for a vital, common purpose. It was a defining moment in my life that gave me hope that we might not be too late to save this very unique planet.

I had a similar experience in June 2020, in the midst of the COVID-19 pandemic. Mass demonstrations were erupting across America to protest police brutality and draw attention to racial injustice. As Tim and I were sheltering in place in San Diego during the pandemic, we participated in a car parade that brandished slogans like "No justice, no peace" and "Black lives matter." The collective voice of so many people saying the same thing, even if we weren't all in the same place at the same time, inspired global citizens to join the fight for racial equity and justice. These events, along with the increasing number of political efforts to reform policing, affirmed my faith in the power of the collective.

The healing power of inclusive group rituals is especially needed during challenging times. Today, with the increasing concerns over the impact of climate change on Earth's fragile ecosystems, I can't help but be drawn to further group action to save this planet and all of its amazing inhabitants. It makes me want to don my mask again,

group rituals

jump overboard into a school of anchovies, and watch the push-pull of nature unfold before me again.

We need to constantly remind ourselves of the importance and urgency of uniting as a group to protect one another and to protect other animals and the natural world. Modern technology has caused many of these problems, but it is also possible for us to apply that technology to restore nature and ourselves. But it is up to us to coordinate and synchronize ourselves as a group to make sure that happens.

3 · COURTSHIP RITUALS

OUTLANDISH ATTRACTIONS

> "Keep some room in your heart for the unimaginable."
> —MARY OLIVER

AS FAR AS THE EYE could see, blue water and sky bent around me with no distinguishable horizon. Only the tufted clouds, reflecting on the ocean, provided a line of sight.

I was standing in the middle of a giant salt pond on the island of Anegada, in the British Virgin Islands, feeling overwhelmed by the vastness of this immense body of water. This low-lying coral island with its secluded beaches and shallow water is one of the most beautiful places on Earth, yet its existence is a paradox.

It's one of nature's flattest landmasses and has an extensive shallow-reef system submerged in turquoise tropical waters, but one could easily be beguiled into thinking that no harm would ever come to pass in such an innocuous environment. The hypnotic blue of the surf, the white sand beaches, and the crystal-clear water are intoxicating. Although it's the second-largest of the British Virgin Islands, it hosts fewer than three hundred human inhabitants, lending to the secluded ambience of this tropical cay.

Yet there's an ominous pall hanging above this oasis—a sense of foreboding. If one isn't prepared for the intense exposure to the elements and the deceptive pitfalls of the island, one could disappear into a salt pond without anyone ever knowing. The last words of lament on one's lips would be Coleridge's famous line in *The Rime of the Ancient Mariner*: "Water, water everywhere / nor any drop to drink." The numerous, calamitous shipwrecks dotting the perimeter of this coral island serve as a reminder of its unforgiving nature.

There I was, with the sun blazing down, scorching my face and arms. The hot mud reeked of rotten eggs, and my feet sank deeper with every step. A cloud of relentless sandflies attacked my exposed, sweating flesh—their bites causing waves of hives all over my body.

I was dizzy with thirst. The putrid mud was now above my shins and rapidly feeling like quicksand. As much as I tried to appreciate my surroundings, I was compelled to focus on finishing my transects to help a friend with her mangrove tree study. The project had already taken many more hours than anticipated. All I could think about was taking a cold shower, drinking a gallon of ice water, and most importantly, never returning to this godforsaken place.

I was startled out of my nightmarish thoughts by the sight of a flurry of pink forms assembling into a line far in the distance. There wasn't a sound for miles but for the strange series of splashing noises emanating from this curious pink entity, which was gathering in size and frenetic energy at an alarming rate.

I stood straining my eyes, trying not to fall over into the algae-rich, salt-pond mud. My friend was so far out of sight that calling to ask her what was happening was useless. Whatever this living wall of pink was, it was moving very quickly in jerky motions in one direction. Then, having reached some seemingly arbitrary point in the middle of the shallows, it turned and marched in the opposite direction.

Momentarily distracted from my plight in the mud flats, and suddenly feeling lighter on my feet, I moved closer toward this curious, animated form. I could now make out a long row of stiff red necks

courtship rituals

with heads and beaks held high. I realized I was seeing something I had only ever witnessed in the pages of a book.

It was March—which is the beginning of mating season for the Caribbean flamingo. What kicks off the season is a very strange ritual—the entire flock of flamingos engages in a group march with synchronized movements. Having conducted a study on flamingo courtship behaviors, I had always hoped to see such a march. My study population consisted of six reintroduced flamingos, however, which was about ten birds less than the critical mass needed to trigger this wondrous spectacle.

Up ahead, about sixty flamingos marched with necks erect and black hooked beaks pointing toward the sky. They flicked their heads back and forth. Accompanying this head-flagging behavior was a specific call that sounded like the honking of geese.

Flamingo group marching usually starts at the end of peak hurricane season and is a very important trigger for a suite of courtship behaviors. These carry on for about a month until breeding pairs form. At that point, nest-building commences. Then eggs are laid, and by mid-April, if all goes well, there are chicks. Flamingos are serially monogamous, mutually choosing a mate for a year. Then come March, they start the ritual all over again with a new mate.

The extraordinary color of the flamingo's reddish-pink plumage also factors into courtship. Flamingos get their fabulous color from eating crustaceans that contain β-carotene—the organic chemical with the reddish-orange pigment that gives carrots and beets their spectacular color. If brine shrimp, copepods, and other crustaceans that feed on micro-algae, which contain β-carotene, are absent from their diet, flamingoes look more white than pink.

Flamingos secrete this pink substance from a preening gland near the tail. Since the labor-intensive application of a flamingo's pink "makeup" only lasts a few days, the color needs to be reapplied every few days. Female flamingos tend to spend more time reapplying than males. Having pinker feathers is associated with access to better-quality feeding sites, which is an attractive quality for a mate in the world of flamingos.

After the group march and rosy attractions, a male flamingo's success at scoring a mate comes down to a prescribed sequence of choreographed postures, executed with precision. The courtship ritual starts off with "head-flagging," where the male stretches his neck high, and with beak up, flicks his head back and forth, accompanied with a specific head-flagging call. The next pose is called the "wing salute," consisting of stretching the neck and head high, while holding their wings out and honking loudly. Next is the "twist preen," tucking neck under wing, and then the "wing-leg stretch," where one wing is held down to the side and the leg is stretched out behind it. Finally comes the "inverted wing salute," stretching the neck straight out and down at the level of the pelvis with wings held out to the sides.

This suite of prescribed steps of the flamingo courtship ritual is repeated over and over again until other males start engaging in the ritual. The increasing number of participants triggers the females to join in as well.

WHILE WATCHING FLAMINGOS engage in this elaborate courtship ritual, it's hard not to wonder how these rituals came about. It turns out that many courtship rituals in birds evolved from stretching and nesting behaviors. This insight certainly seems to make sense of the flamingo ritual: stretching its long leg out behind, holding wings wide to the sides, twisting around and pretending to preen. In some species, however, ritualized movements or displays are strikingly unique and not like any routine behaviors. These specialized behaviors are designed to show off the strength or health of a would-be mate in a quick encounter, rather than enticing others to join in what had once been, perhaps, a relaxing stretch within the colony.

courtship rituals

Either way, males that perform the most desirable or unique displays get to mate and pass their genes to the next generation. After many generations, these behaviors become fixed in the population. So, for many species, courtship rituals are encoded in their genes.

Courtship rituals in flamingos and other colonial breeding birds are similar to human group rituals like ballroom dancing, swing, or salsa. They are most similar to square dancing, in that individuals participating in the dance engage in a series of organized and synchronized actions that result in a pairing off of couples. If one is lucky, this pairing off endures longer than the square-dancing encounter.

In some other species like the blue crane or bald eagle, pair bonding can be for life—so a lot is riding on a female making the right mate choice. Courtship rituals serve as an important opportunity for a female to assess the qualities of a potential mate. In more promiscuous species, males gather and compete for the attentions of a female in what's called a "lek." With stiff competition all around, a male doesn't typically have much time to demonstrate his skills. This is why he has to pack as much as possible into his show—including flashy colors. If the male's stage for his show is the cluttered understory of the jungle, the display has to be more elaborate than if he were displaying on a tree branch.

The more elaborate the display, the quicker a male can gain and keep the attention of the female. If his fancy footwork and feathery displays succeed in getting him a mate, the genes for these traits get passed on to the next generation. If the female is looking for assistance in raising her young, the display takes on a different tenor, as she's also assessing the male's parenting skills.

Courtship rituals can be very costly for males. Their extravagant performances and bright colors can make them more conspicuous and vulnerable to predators. So, why did these behaviors evolve and why would they persist if they are so dangerous?

As Darwin explained it, a male peacock's tail evolved because the success that the male had in mating with such a splendid ornament outweighed the costs of avoiding a predator by being drab. This is how Darwin came up with the idea of sexual selection. Contrary to his theory of natural selection, females choosing particular traits can drive evolution just as much, or even more so, than outside forces, such as weather or food availability. The traits that the female selects in a male get passed on to the next generation.

The indicators of fitness in bird courtship rituals can be a feast for the eyes, ears, and olfactory bulb. Sometimes, they even include a tactile component. Males are put to the test by demonstrating their rigor through song, dance, breeding plumage, and in the case of the bower bird, even their ingenuity and artistic flair.

Often the color red is a signal of high testosterone in males—such as the red comb on the rooster's head or the pink beak and shins on the male ostrich during mating season. It is energetically expensive to produce testosterone, so these displays are an honest way to signal fitness to the female during courtship. The same is true for musth in male elephants. Instead of using the color red, however, they generate a scent and display a whole lot of swagger to advertise their fitness, as it takes a lot of testosterone to maintain the hormonal state of musth.

To impress a female, many male birds combine outlandish dance moves, elaborate plumage, and sometimes a serenade, and of course, a lot of repetition. The red-capped manakin moon walk is a particularly stunning feat of footwork across an open branch stage in the forests of Central and South America. The male hooded grebe incorporates a head-banging behavior not unlike heavy-metal concertgoers. Even more amazing is the western grebe mating dance that includes the couple literally walking on water. By rapidly stamping their feet at fourteen to twenty steps per second, they are able to rise out of the water and sprint across it, covering sixty feet in a few seconds.

courtship rituals

An ostrich mating dance is another bizarre spectacle whereby the male presents his black-and-white feathers by waving his wings up and down in an undulating pattern, while collapsing his neck, with pink beak opened, and twisting his neck like a snake across his body. Since this ritual culminates in a seated position, with wings swishing the sand back and forth, before jumping onto the seated female's back to copulate, this elaborate dance ends up looking more like nest-building than a courtship ritual. Interestingly enough, this last step in the ritual is, in fact, thought to have evolved from nesting behaviors.

The most elaborate male bird courtship rituals are best showcased in the bird of paradise species in New Guinea. The extraordinary displays performed by these birds—employing postures that produce the most surprising feather arrangements and color combinations—are truly a marvel of nature. It is hard to imagine that this group of birds evolved from drably colored crow-like ancestors twenty-four million years ago.

Amusingly, female birds of paradise scrutinize the elaborate performances of their colorful male suitors with apparent indifference. Their discerning eye is a prime example of how sexual selection drives the evolution of the most extraordinary behaviors in nature.

Some of these bird of paradise displays rely on more than just feathers and dance moves. Take the male Wahnes's parotia, for example—one of those promiscuous species that have to work especially fast, and with perfection, to succeed in mating. He looks for a patch of ground in the forest where a female could watch his performance from a branch above. Then he sets about clearing his stage of leaves and debris. When the area is immaculate, and his audience has arrived, he begins his dance for the female by shaking his head back and forth and fluffing out his pitch-black feathers into a tutu.

Above him, the female sees an iridescent blue breastplate in the center of the oval tutu as well as a patch on the back of his head—the effect looking like a tribal mask shimmying back and forth. This elaborate display seems like it would mesmerize any female being courted, so much so that one has to wonder what criteria a female uses to judge a male's performance.

The courtship display of the male golden-collared manakin of Columbia and Panama has been studied well enough for researchers to figure out exactly what the female is assessing from the male's ritual. The physically elaborate, and hormonally costly, displays of these birds are quick, precise, and powerful. After much scrutiny of which displays were preferred by females, researchers determined that females choose males based on their motor skills. Since the neuromuscular specializations required to perform the ritual are energetically expensive, this tells the female that the male is a suitable mate.

A recent study of the houbara bustard shows that there is indeed a lot going on in that female bird brain while she takes in a bit of courtship eye candy. Merely watching the display can lead to higher fertility in the female and even healthier chicks. The chicks are healthier due to the mother incorporating additional compounds within her eggs that lead to an improved growth rate in the chicks. This finding adds a new dimension to voyeurism—and takes the idea of having a bridesmaid party at Chippendales to a whole new level.

Flamingos take a different approach to watching. After witnessing the males' display, females are triggered to fully participate in the courtship ritual themselves. This is true for other bird species like the blue-footed booby on the Galapagos Islands. The male booby presents the female with his lovely blue feet by prancing around her in exaggerated steps. If she chooses to join in the ritual, this escalates into a bow and stretch that they both perform to each other. Then, at some point, the male will offer the female the nuptial gift of a twig. It's not just his blue suede shoes that are on display. Male and female boobies both assess the blue quality of their mate's feet in courtship (a signal of their health), because both are invested in parental care of the nest.

courtship rituals

The great bowerbird of northern Australia has an entirely different strategy. Rather than theatrical acts or flashy colors, the male bird constructs a bower, a work of art that he presents to his prospective mate in the hope that she will find the artwork attractive enough to mate with him. If the bower doesn't get sabotaged by the competition, the male will also sing to her. If she is receptive, they will dance and mate. After which the female disappears, and he starts the process all over again.

Like the bowerbird, a male human can create a dazzling array of props to improve his prospects. These props are referred to as secondary sexual traits, as they are used to attract a mate. Studies have shown that both men and women tend to incorporate sex-specific products within courtship rituals, where women may focus on beauty aids (the bird of paradise strategy), and men may focus on objects such as sports cars (the bowerbird effect).

A recent experiment showed that these sex-specific products can impact our hormone levels. For example, a man's testosterone level was shown to be elevated while driving a Porsche versus a regular sedan. Another study showed that having possession of a sports car influenced both men's and women's perceptions of other men's physical features. When women were presented with a picture of a man standing in front of a red Porsche for an online dating site, the women automatically assumed that the man associated with the high-status product was tall. Men presented with the same photo automatically assumed that the man was short, demonstrating inherent male-male rivalry.

Another study on women's behavior during their ovulation cycle found that women spend more time on beautifying efforts during the window of ovulation. This suggests that our subconscious is more aware of our hormonal state than we might anticipate.

Courtship in elephants, on the other hand, is a much less colorful affair. Far more effort goes into olfaction and acoustics. To draw an estrus female and musth bull together from miles apart, elephants use scent and specific courtship vocalizations to find each other. Since the gestation time for an elephant is almost two years (twenty-two months), and a calf nurses for another two years, a mature female only comes into estrus once every four to six years, making it quite a rare event and all the more important to seize that few-day window when it arises.

During this time, a female emits a long low rumble vocalization that she repeats frequently. This longer repeated rumble vocalization provides the biggest cue of her status—along with the hormones contained in her urine. The same is true for the musth rumble. Once a musth bull is in the vicinity of an estrus female or any perceived competition, the displays begin. He wafts his scent into the distance by waving his ears forward, one at a time. Then he rubs the secretions from his temporal glands with his trunk and flings them into the air. All the while, he dribbles a trail of urine loaded with his scent. All these acts serve to both distribute his scent and intimidate other males.

Humans use scent in their courtship rituals, too. Similar to female and male elephants, women use perfume and men use cologne to advertise themselves.

In one courtship ritual in Austria, young women perform a ritual dance with slices of apple held in their armpits. After dancing, the women give the apple slices to the men of their choice, and the men eat them. This ritual might seem strange, but it's just another way we use smell to choose a mate—similar to the pheromones in insects, musth in elephants, or perfume and cologne in humans.

Studies show that women unconsciously choose a mate based on a man's scent. In one study, a group of women were presented with a selection of T-shirts that had been slept in by different men who had varying body odor, ranging from familiar to foreign. The women chose shirts that smelled familiar, but not too familiar. This result has evolutionary underpinnings: Humans can discriminate relatives based on scent, and this ability would prevent inbreeding, just as it does in many species that otherwise have no information on relatedness, which happens when species don't live socially as adults.

courtship rituals

ALTHOUGH MALE-MALE COMPETITION is at play in the lion kingdom, it's the lioness that uses courtship rituals to keep her lover on schedule. The lioness is not a silent partner in this affair, which won't be surprising to anyone who has had the experience of being kept up all night by the neighborhood female cat during courtship.

Just like domestic cats, lions need to copulate in order to induce ovulation—and usually multiple times, sometimes on the order of a thousand times (according to one report)—so it could be said that lions are not lazy lovers. The nights can get very long indeed for both lions and nearby researchers during these all-night mating episodes right next to camp.

This daunting number of copulations explains why the lioness is particularly motivated and uses all of her charms to keep their carnal activities on an ambitious schedule. With a purr and a saunter, the lioness walks past the male. As she does, she draws her tail across the front of his face, ending in a face full of tail tuft. This action appears to be irresistible no matter how flat-out exhausted the lion is.

The presence of courtship rituals in many nonhuman primate societies is dictated by whether mate choice is exhibited or even possible in any particular species. Mating opportunities are often determined by status within the group. For example, female primates tend to prefer the dominant male. This means that a male with lower status may have no chance for a mating opportunity until they are fit enough to take on the dominant male. Then again, in other cases, like in vervet monkeys, only higher-ranking females are able to reject advances of low-ranking males, so female choice is sometimes confined to females with higher status.

In baboons, males are twice the size of females, so female choice does not play much of a role, or at least it is harder to measure. When female chimpanzees court the dominant male, it is an extremely practical affair: She simply presents her bright red backside to the dominant male to let him know that she is in estrus. Since a male only mates once a day, if the female is rejected, there is a lot of opportunity for subordinates to step in—and thus good reason for lower-ranking males to court females with such behaviors as helping to carry a baby. Male chimps know how to make up for the lack of rank or brawn to impress a female with generosity and kindness.

Pair-bonded (monogamous) primates exhibit courtship rituals such as mutual gaze and grooming. This grooming behavior is similar to brushing or playing with a partner's hair, which releases oxytocin—or what's referred to as the bonding hormone. For both monogamous humans and nonhuman primates, such as the titi monkey, long-term relationships produce "feel-good" chemicals like dopamine and other beneficial chemicals that minimize stress and enhance group cohesion.

BIRDS AREN'T THE only animals with wildly varying courtship rituals shaped by geography, environment, and status. Human rituals vary, too. Visit the desert, jungle, or savanna, and you will find people adapting their courtship rituals to the places and societies where they live. For example, the sing-sing festival in Papua New Guinea is an annual gathering of villages to share their traditions. Each culture has its own signature sing-sing, partly inspired by the courtship rituals of birds found in local forests. Men dress up in elaborate costumes, incorporating a particular bird coloration and feathers as well as its song into a courtship dance to attract a potential lover. Other cultures, in addition to adopting elements of bird courtship, also include fancy dance moves like the tango or Spanish flamenco.

Most human courtship rituals have the same goal as those of other animals: They provide opportunities for a woman to assess a man's attributes outside of sex. One example is the serenade or duet, whether it involves dance, poetry, song, or playing musical instruments. Performing alone or together in any of these ways can be an expression of interest or commitment.

courtship rituals

Recently, my husband, Tim, and I attended a piano recital at the Salk Institute, where the renowned pianist couple Alessio Bax and Lucille Chung played a two-person piece together. Their performance was billed as "four hands, one heart," which I thought was particularly apt. They played with such passion that Tim remarked it felt like we were voyeurs in a transcendent sexual encounter. Their intensity caught me off guard. The sheer power of the experience brought me to tears.

As I walked out of the auditorium, wiping my tears, I wondered how two people might generate such an elevated courtship experience without being able to wield this powerful instrument together. Then I realized that, even if Tim and I were just watching, *our shared experience of such beauty was a form of courtship*. No matter how new or mature a relationship is, dating plays an important role in creating a point of connection and intimacy because it offers the opportunity to have a shared experience.

How does a human courtship traditionally begin? With an expression of interest. In the United States, this is enshrined in Valentine's Day, with its gifts of chocolates, flowers, and jewelry, and the secret or not-so-secret ritual request: "Will you be my Valentine?" This modern courtship ritual evolved from an ancient Roman festival that marked the official start of spring, sometime in mid-February, and dates back to the year 496.

A more peculiar courtship ritual called "bundling" dates back to biblical times. It was once practiced in the Netherlands and is still practiced today in Pennsylvania Dutch country among some Amish cultures. In the bundling ritual, a young teenage boy and girl stay overnight together at the girl's residence, and they are bundled in separate blankets within the same bed, and further separated by a "bundling board." Despite being wrapped in separate blankets, the couple has the opportunity to experience intimacy without having intercourse—and it is subject to parental oversight as additional insurance that nothing more than words and breath are exchanged.

The Dai culture of southern China has an annual courtship ritual called "visiting girls." Young women sit around a bonfire turning their spinning wheels, and they are visited by men wearing red blankets and playing instruments. Each man chooses a woman to serenade, and if she likes him, she invites him to sit on a small stool that she pulls from under her skirt. The man then wraps her in his red blanket and they whisper to each other.

Time also plays a role in courtship rituals. Longer courtships can incorporate symbols to reflect the amount of time the courtship has taken place. A ring might be worn to represent "going steady" or getting engaged. Sexual intimacy is often seen as a measure of seriousness. However, one study found that relationships are much less formal today, and couples said they don't need symbolic gestures to reflect a level of commitment.

Courtship rituals aren't only about starting something new, either. They provide fresh and exciting ways to rekindle an old flame or restoke a current flame. As I was writing this book, I recognized that my husband and I have been remiss in maintaining simple but important courtship rituals, such as gaze and touch. Like nearly everyone else, we enjoy binge-watching Netflix—but we also enjoy meaningful conversations, holding hands during a walk, and making eye contact and smiling during dinner. Sometimes, however, even the simplest gestures can get overlooked. Yet, even small gestures help strengthen our connection. As I've mentioned, even simple gestures like smiling and laughing are contagious behaviors. They reinforce relationships as well as improve our own health by reducing stress—especially when we direct this positive energy toward a loved one.

On the other hand, you could say that every ritual of courtship does start something new. Even if it is something we do every year or every day—it is never the same experience. Done with intent, courtship gestures can have the same thrill as when done for the first time with a new romantic interest. This is the goal when married couples renew their vows. Partners can adopt any activity and make it a ritual that reaffirms their love and commitment.

Consider what you might do in your life, either within a current relationship or to attract the romantic interest of someone else. Bowerbirds and flamingos don't wait till potential mates are watching to begin their rituals. Their activity draws attention and then aims to keep it. This could be something that you do with a partner, a group, or even by yourself. Start something, create a habit, and see if you attract others to your ritual.

For instance, you could initiate a dinner club or book club—or a club related to whatever interest you find rewarding. You can get involved in some new activity as a couple, whether active and outdoors or relaxing and indoors, and learn something new together. Don't wait for others to start something and invite you to join. Be proactive.

Beginning something new can also be contagious. It can help us see other areas of our life that need refreshing. Courtship-type rituals can lead to new friendships in addition to courting love interests. And of course, courting a significant other can also entail courting their friends and family and expanding your relationships in many unexpected ways, including making plans to expand your own family.

In thinking about our similarities with our great ape kin, I have to say that I am glad that women don't have to advertise their ovulation window with an engorged red rump like chimpanzees. Having a potential suitor eat apple slices from under a woman's armpits doesn't sound all that tantalizing, either—but to me, it is definitely preferable to bundling. Still, even in modern human courtships, some ritual is fun, even if it's not as elaborate as a flamingo group march or taking up flamenco dance. All social creatures go a-courting in some fashion. It's how we attract the relationships that are ultimately essential to perpetuating our species.

4 · GIFTING RITUALS

SHINY OBJECTS, FLOWERS, AND DEAD BIRDS

> "I have found that among its other benefits,
> giving liberates the soul of the giver."
> —MAYA ANGELOU

DIEGO, A 110-YEAR-OLD GIANT TORTOISE, is credited with single-handedly repopulating almost half of his endangered species on the Galapagos island of Española. In 1977, he began sowing his seed in captivity in Galapagos National Park, and in January 2020, he retired and returned to Española after having carried out one of the longest recorded courtships in nature and siring at least nine hundred offspring.

Diego's courtship ritual featured a very important gift—a wild Galapagos tomato, a favorite food of the giant tortoise. Diego was a very reliable gift-giver. Each day, he would diligently and very slowly stroll up to his desired mate and drop a small yellow tomato at her feet. Then one day, his gift-giving stopped.

No one will ever know what made Diego cease his long tradition of courting and proffering tomatoes, but Diego is not the only social animal to use gifting as a means of communicating. This ancient ritual of a male offering a female a *nuptial gift* is common throughout the animal kingdom, including among insects, birds, squid, dolphins, monkeys, and great apes. Many animals present such gifts to "sweeten the deal," enticing a mate to choose them over the other suitors. In the case of food offerings, this also provides a nutritious snack, enabling their mate to lay the healthiest eggs possible.

I was once the recipient of a nuptial gift from a very handsome southern ground hornbill named Gumby. With eyes and long feathery eyelashes that looked disturbingly human, Gumby befriended me at Zoo Atlanta with the gift of a decomposing dead bird. Unusual as it might seem, sometimes nuptial gifts can be presented to other species, outside their intended purpose. As I watched the behavior, it struck me that there is often an urgency surrounding this kind of gifting in nature.

Gumby and I had been introduced when his keeper took me behind the scenes into his exhibit. In seconds, this black-and-red carnivorous bird was beside me, nudging his head in my direction, while clutching his prized dead bird within his large, curved beak. I was now his new crush—the researcher who had come to learn about his courtship habits.

This insistent suitor acted as if his life depended on my acceptance of this very odd gift. He completely ignored his attractive female companion, Zazu, but she seemed nonplussed by this new development and stood quietly next to him. She appeared quite content to be relieved of his incessant courtship overtures.

Even after I left the exhibit, Gumby continued pacing and tapping at the glass, still hell-bent on depositing his gift with me, which dangled from his long, black beak. He kept this racket up, in and around a tangle of other zoo guests, to make sure he was keeping my full attention. Granted, he was handsome with his bright red face and red throat sac—but no one finds a desperate suitor attractive. To be honest, his persistence was starting to give me the creeps.

But not Zazu, who continued to stand calmly on the branch right next to him, unfazed by his crazy behavior and acting as if everything was normal.

The urgency of giving a nuptial gift and having it be accepted is understandable from an evolutionary standpoint. If a female doesn't

gifting rituals

accept the nuptial gift of a male, she is most likely not going to mate with him. Depending on how many other receptive females are available, this could mean that the male doesn't succeed in passing on his genes. Or it could mean that he has to wait for the next breeding season, *if* he is lucky enough to live that long.

Many insect species have a very short adult life, so mating actually could be a once-in-a-lifetime occurrence. A creature might get only one shot to pass their genes on to the next generation to carry on their genetic line.

Whether it be a wild Galapagos tomato or a dead bird, nuptial gift-giving in nature can have many meanings. Take our overeager friend, Gumby the hornbill. Just by being fit enough to obtain a gift, he signals to the female he has healthy genes to pass on. A nuptial gift can also directly contribute to the female's sustenance by offsetting the energetic cost of nest-building, egg-laying, or having to nurse or feed offspring. Often, the gift can be symbolic. In the case of the blue-footed booby, the twig or stone that he offers within his courtship ritual is merely a prop, since these birds don't actually build nests. Blue-footed booby eggs are laid on the bare ground. The symbolic gift of nesting materials is thought to be a remnant of a behavior from an ancestral species that used twigs and stones to build nests, and over evolutionary time, environmental pressures were such that nests were no longer built and nesting behavior was reduced to these token ritual gestures.

THROUGHOUT NATURE, in human and nonhuman animal societies, a gift is a form of communication, and it has the ability to modify or completely change a relationship. Gifts are also given to reinforce an existing relationship or begin a new one. This isn't to say we can buy love through objects, but what we give others has a powerful impact on our relationships. A gift can offer protection or reciprocity or even be a method of socialization, like giving the new neighbor an apple pie as a welcome to the new community.

In its simplest form, we can think of a gift as a tangible expression of the status of a relationship in the moment. Giving too much or too little, or too late, can lead to negative consequences, which is why gifting rituals are replete with decorum.

Our earliest ancestors used gifts to demonstrate a man's ability to provide, which improved his chances of finding a mate and creating a family. Even our closest evolutionary cousin, the chimpanzee, exchanges gifts. Male chimps will often offer food gifts, such as highly prized meat, to females, in the hopes of mating with them. Or they might offer favors, such as a grooming session. Modern humans offer food gifts, too, in the form of a romantic dinner with a bottle of red wine.

Although the offer of something to drink to a guest seems ubiquitous in modern cultures, the custom has been around for a very long time. In ancient Mayan culture, a host was expected to give something to their guest, even just a drink of water. In context, this typical Mayan gift makes sense. The Mayan city of Tikal used sustainable water-management technology to support tens of thousands of people for over fifteen hundred years.

Today, our gifts are often symbolic offerings, such as a bouquet of flowers or a ring in a little black box, in the hope of establishing a relationship or strengthening one. This is not unlike the bowerbird presenting a piece of artwork, in the form of a woven structure, as a mating contract.

Even space can be a gift. The meaningful offering of an area in our homes to our significant other, like room in our closet, can signify a new step in a relationship—one that introduces the possibility of coexisting.

The human custom of giving financial gifts to a bride and groom, to help set them up in their new life, is a long-standing tradition. Dowries are still expected in some cultures, whereby the family of the bride offers a financial gift or a number of head of cattle to the groom's family.

For humans, the ritual of gifting has endured throughout our evolution in various forms for good reason. Despite culturally distinct practices, the underlying motivations and elements are consistent. The act of gifting is a token of appreciation, of love and friendship, and a form of remediation. Gifts are also representations of power or sharing one's power, and they can often contain the expectation of obligation.

There are two kinds of gifts: *transactional gifts* and *reciprocal gifts*. Transactional gifts are given with no expectation of a return gift. A man giving flowers to a woman as an expression of love doesn't expect her to give him a purchased gift in return. However, there is an implied transaction—some benefit or reward is expected, either conferred in the moment or at some point in the future, no matter the intention behind the offered gift. For example, by accepting a gift of a piece of jewelry within a romantic relationship, this often extends the promise of further commitment.

The same is true in nature. Among nonhuman animals, nuptial gifts come with the expectation of mating. Sharing food is also transactional and can be a form of parental investment. Food sharing can enhance one's fitness and help to ensure the survival of one's offspring. For example, a lioness might allow her adult daughter to steal from her kill, and by sharing, the daughter could help the lioness defend a food source in the future or help her in her hunt.

Beyond the interpersonal, transactional gifts are common in modern human society. Take philanthropic gifting, which usually comes with a token return depending on the size of the gift—such as a tote bag for donating a small amount to your local NPR station, or a free round of golf at a fancy club for a much larger donation. A true philanthropic gift does not have any strings attached, although social psychologists argue that "there is no such thing as a free gift," meaning that all transactional gifts are ultimately reciprocal in nature. After all, even Santa Claus expects you to be "nice" in order to receive a gift.

Some of the most famous transactional aesthetic gifts were made in the context of love. The Hanging Gardens of Babylon are considered one of the seven wonders of the ancient world. Although there are conflicting reports as to whether the gardens actually existed, written accounts suggest that King Nebuchadnezzar II had this highly impractical gift built in the sixth century BC to remind his wife of her home in Persia (modern-day Iran).

The Taj Mahal was built in Agra, India, as a shrine for Mumtaz Mahal—the wife of the fifth Mughal emperor, Shah Jahan, in the middle of the sixteenth century. Another example is the first Fabergé egg, which was given in 1885 by Tsar Alexander III to Empress Maria Feodorovna as an Easter present, and perhaps also to honor their wedding anniversary.

Gifts from one nation to another, in order to foster peace and mutual understanding, have origins going back to at least the time of ancient Egypt, with a practice of providing a neighbor nation with stone jars inscribed with a royal monogram. A modern example is the Statue of Liberty. Gifted to the United States from France, the statue commemorates our alliance during the American Revolution, as well as our mutual commitment to freedom and democracy.

Reciprocal gifts refer to when related or unrelated individuals offer each other something for their mutual benefit, a gift for a gift. Sometimes the expectation is overt and sometimes implied. For instance, a lioness sharing her kill with an unrelated female within the pride might do so to avoid a potential contest or fight over the food. In this way, the clever lioness avoids the possibility of injury and of getting her meal completely stolen. This act may also improve the unrelated female's chances of survival and her likelihood of sticking around. Sharing a kill can be a small sacrifice or gift that allows the lioness to expand the pride and help all the females defend their cubs from infanticidal males.

Food sharing has been demonstrated in such diverse species as nonhuman primates, elephants, lions, wolves, rats, birds. and bats, showing that gifting is a ritual of survival throughout the animal kingdom. This act builds alliances and trust between individuals and within a group.

The notion that "you scratch my back, I scratch yours" is very common in nature. In one fascinating study, researchers explored helping behavior among birds in the context of food-sharing. Captive African gray parrots were given a valuable token that could be exchanged for a precious walnut treat. These parrots voluntarily and spontaneously offered a token—in some cases, up to ten tokens—to another gray parrot without receiving anything in return.

At a later time, the parrots that received this gift were given a token to see if they might return the favor. They did not disappoint. The parrots reciprocated the gift, while watching their partner receive a walnut treat. The strength of the relationship between the parrots affected the willingness to share, which implies a previously established understanding of reciprocity. In contrast, when blue-headed macaws were presented with the same challenge, the test subjects were not inclined to give up their walnut tokens under any circumstance.

A similar study conducted with three-year-old children found that resource sharing with a partner was more generous if the partners had previously shared with each other. Researchers suggest that reciprocity might also be an expression of gratitude. Children as young as a year old are motivated to help others, suggesting that helping in humans is instinctual.

While some acts of giving involve delayed benefits, many are for immediate rewards. In nature, one such behavior is "recruitment calling" to share food. This is very common in birds, whereby an individual alerts others to food in the area. In predators, recruitment calling includes the "whoop" call among hyenas, which serves to gather the clan to share food, with the expectation that a larger group would help defend a kill from lions. Cliff swallows commonly announce a food source such as a swarm of insects by giving a squeak call. This sharing leads to a greater haul for the group and expands opportunities to find food. However, the larger the group of birds, the more predators they attract, so sharing has its risks.

The house sparrow has a solution for the problem of attracting predators. They will gauge the amount of food that is available and recruit others if there is enough food to share. As the flock size increases, the rate of recruitment calling decreases.

Whether altruistic or selfishly motivated, helping others *actually benefits the giver*. Killer whales, for example, benefit by sharing a carcass of a seal because multiple whales feeding on the seal will keep it afloat for longer than if a single individual were feeding on it.

In order to avoid being lunch during sex, the male praying mantis offers a food gift in order to distract the female from cannibalizing him. Many animals invite others to share in the food they have gathered or hunted in order to reduce the possibility of getting eaten by a predator while they are distracted by eating—more eyes on the lookout.

Another reason many animals share is to mitigate harassment. Studies have shown that an individual in possession of food will get harassed by others, causing its stress levels to increase. By sharing, it can avoid potential injury or further harassment. Outside of kin or associates, chimpanzees share their food with those that harass them four times more often than with those that don't.

Gifts can also serve to reduce stress in the context of a social hierarchy. Gifting down to a lower-ranking member of the group is common among birds such as the babbler and the raven, and among chimpanzees. In the babbler, there is a strict dominance hierarchy. Gifting of food only occurs from the dominant male to the next-ranking male, and the acceptance of the food decreases the status of the recipient. According to recent baboon research, the alpha male is more stressed than the beta male in this situation, so this kind of gifting might serve to mediate stress in some way.

Among early humans and in hunter-gatherer societies, resource sharing provides insurance when food security is uncertain. For example, sharing a particularly abundant harvest or food cache encourages others to do the same in the future if one's own harvest fails or if one is unable to find enough food to sustain one's family. This same dynamic exists among vampire bats. If one bat is unsuccessful at obtaining a blood meal for two nights in a row, it will die unless another roost-mate gifts it with a regurgitated serving of blood.

MANY HUMAN CULTURES have different gift-giving rituals. Social psychologists have tried to understand gifting from an evolutionary standpoint by focusing on the process rather than the event of presenting the gift. Often, the gift-giving ritual has been viewed as a social obligation. At times, it can cause the recipient of a gift a significant amount of anxiety, since people can feel both a need to repay the giver and to offer up something of equal value. This is what social anthropologists have called the "gift-giving paradox."

Take, for example, the Chinese custom of giving either money or cigarettes. Gifts in China have to represent a value that honors the relationship between giver and receiver. When meeting an old friend or business associate, gifts are expected, even if the giver is not in a position to afford to buy them. It's hard to imagine that cigarettes are still a currency of gifting, considering their health dangers. Nonetheless, cigarettes are an important part of the gifting culture in China.

When I was in southern Yunnan interviewing local leaders about elephant conservation issues, I experienced the tradition of cigarette gift-giving firsthand. My Mandarin interpreter's old friend met us at the airport in Kunming—the capital of Yunnan. He gave her two boxes of very fine cigarettes, as the best tobacco is grown in Yunnan, and then he presented both of us with a box of fine green tea from the region. I immediately worried that we did not have any gifts for him and asked my interpreter if this was a problem. She instructed me to pay the dinner bill as the reciprocal gift, which I happily did. In this case, the gift-giving paradox was not an inconvenience.

There are other contexts for gift giving, such as giving a gift to oneself. Considered a form of self-care, self-gifting is now common in Western societies. It can either protect one's self-esteem or reward oneself. Buying new workout gear, for example, can be a powerful motivator for achieving an exercise milestone. In Eastern cultures, researchers believe that self-gifting may exist as a material means of obtaining an ideal self or as a strategy to feel more fulfilled.

Another form of gifting is the peace offering, whereby the giver is asking forgiveness for some past transgression. Male elephants do this by placing their trunk in another's mouth, in a trunk-to-mouth exchange, as a form of reconciling after a previous altercation. If two chimpanzees have an altercation, they come together afterward and offer a hug and a kiss on the lips. In bonobos, researchers believe that the gift of a sexual encounter reduces tensions surrounding food sharing—which gives a whole new context to "friends with benefits."

Western society, as a whole, has expressed gift-giving fatigue in the last few decades. All the commercialization that has sprung up around the winter holidays has spiraled out of control. With online organized options, such as wedding and baby-shower registries, all the practical aspects of giving are managed. This means we no longer risk receiving ten toasters at our wedding, nine of which need to be returned. Yet this risks taking some of the meaning and benefit out of giving.

Commercialization can hijack our rituals. In response, some married couples ask their wedding guests to forego presents and make a

gifting rituals

donation to a charity instead. This rejection of commercialization is noble and yet some opportunity to create a further bond between the giver and receiver may still be desired.

When gifts are personal, and symbolize our affection and knowledge of others, it can strengthen the relationship. A gift shows our appreciation in a lasting way, as a reminder that inspires continued connection. Just as we bear witness to a couple's marriage, a gift can be an investment in their bond and in our relationship to the couple that extends far beyond the wedding day. For instance, a gift of personalized lessons related to a skill we possess—like with a musical instrument or as a chef—creates opportunities for new shared experiences.

The goals of any gift—whether it's to commemorate a wedding, a birthday, a graduation, or even just to lift a friend's spirits when they are having a bad day—are to be a reminder of our relationships, to express our appreciation to someone by showing that we are thinking about them, and to solidify our commitment to that relationship and its future.

There is a reason gifting as a ritual has persisted throughout the ages. Giving is more about the giver than the receiver. The desire to be remembered plays an important role in the gifting ritual. My parents have many heirlooms that were given to them as wedding presents by relatives. Through these heirlooms, the giver is remembered in perpetuity, every time the gift is used. The memory of the givers and their relationship to the recipient is what is most important. For recipients, acknowledging and appreciating a gift is also an important aspect of the gift-giving ritual.

Why? Because providing pleasure for someone else is another reason we give gifts. For example, a recent UK study of couples taking a three-week massage course found that giving your partner a massage can be just as rewarding as receiving a massage. Those who gave the massage reported just as much of a reduction in stress as those who received the massage. The giver reported a significant increase in physical and emotional well-being, purely by providing a pleasurable experience to the recipient.

This is also true of pet owners. Lavishing attention on a dog, in the form of belly rubs and ear scratches, and seeing the pleasure the dog experiences, is hugely rewarding. The same is true of tickling a baby. Attention can also be considered a gift—particularly when the time and attention is significant, such as when volunteering at an animal shelter or a seniors' facility.

When my dog presents my husband with his favorite squeaky toy gift or me with a live mouse, I have no doubt that he is enjoying the rewards of a gifter. Presenting offspring with a prey item as a practical gift to teach them hunting skills is very common in nature. For example, meerkats eat scorpions and have to learn how to eat them without getting stung. A pup is first presented with a scorpion with the stinger removed. After some trial and error, the pup eventually learns to remove the stinger itself and thus becomes able to capture its own scorpions.

Similarly, a lioness will often present her cubs with wounded prey as a gift for them to play with in order to learn how to hunt. Bobtail, the resident lioness at our elephant field site in Namibia, found an irresistible gift for her cubs one day when she stole our very expensive microphone and black fuzzy windsock—mistaking it for a dead animal. It was protected under a pile of thorn bushes near the waterhole, the bushes placed there to keep such an intruder at a safe distance from our sound-recording site. On that day, we had left it semi-exposed after replacing a flat battery. Bobtail lay on her belly, reached her paw through the thorns, and snagged it.

Bobtail grabbed the windsock in her teeth and ran off with her cubs trailing after her. At the other end of the clearing, she presented her special gift to them. Her cubs promptly batted it around the waterhole until Tim arrived in the truck to retrieve it.

gifting rituals

When Bobtail saw Tim coming, she grabbed the windsock and ran off into the dense bush. This was too good of a gift for her cubs to want to return it to us. Fortunately, Tim's pursuit of Bobtail into the bush with the truck must have made her nervous. Realizing her cubs had been left exposed in the clearing, she finally dropped the windsock and ran back to them.

As we watched the lions disappear into the bush, we were extremely grateful that Bobtail had given up her gift. We made sure to pile on extra thorn branches so that she wasn't tempted to steal our microphone again. This episode not only highlights the risks one might take to obtain the perfect gift, but also a statement about cats, both wild and domestic. Maybe a black fluffy toy is simply so irresistible that a lioness will risk the security of her cubs to get it, gift it, play with it, and fight for it.

5 · SPOKEN RITUALS

RUMBLING, ROARING, AND YODELING

> "There is a language going on out there—the language of the wild. Roars, snorts, trumpets, squeals, whoops and chirps have meaning derived over eons of expression.... We have yet to become fluent in the language—and music—of the wild."
> —BOYD NORTON

WAS DRINKING TEA AND ENJOYING the pink late-afternoon glow over the savanna. The air was still, as it often is at this time of day. Sitting high above the ground in the research tower at my field site in Namibia's Etosha National Park, I watched as Ursula, the matriarch of an elephant family we had named the Goddesses, took a leisurely mud bath with the rest of her family in the shallow watering hole.

All elephants love a good mud bath, but Ursula was particularly fond of them. She luxuriated, sucking up a trunkful of mud, swinging her trunk around her body, and releasing the mud with a loud slap against each flank. This was time to herself, without having to be concerned with the safety and politics of family.

Then I noticed a thick trail of white dust hanging atop the tree line in the distance. An almost imperceptible noise followed, like the hiss of an approaching rainstorm. The noise got louder and louder, transforming into a thunderous roar, as a very large group of elephants broke the cover of the forest's edge and stampeded into the clearing.

In the center of the group of over forty elephants, all running at full speed toward the water, was Chilonis, matriarch of the Scimitar family. Her ears were taut and pinched aggressively, and her imposing, wide-splayed tusks looked sharpened for battle. She waved her massive, pale gray head back and forth in a threatening ceremonial gesture.

Then a member of Chilonis's family let out a blood-curdling cry, and the call to arms was followed by roaring and bellowing from different family members as they continued advancing. Their intentions and message were unmistakable—to drive Ursula and the Goddesses away and lay claim to the waterhole. They were prepared for war if necessary.

I never know what to expect in a field season from one year to the next. More or less rain can dramatically change the elephant dynamics at the waterhole. This season followed the worst drought in ninety years, so access to a waterhole represented a family's lifeblood. Without water, an elephant family can't survive—and with so little to go around, staking a claim was critical.

Only recently, Chilonis had been pregnant and couldn't move faster than a waddle. But three days ago, she had given birth. Now she had a newborn at her heels, scampering to keep pace, and nothing was going to slow her down. The battle cry from her family made it clear she would fight for this precious resource.

Chilonis and her Scimitar family had a fairly recent history at Mushara. During the previous El Niño cycle, a number of floods had stranded family groups in the northeast that were typically from the west of the park. Suddenly, even more families were competing for the region's very few waterholes, and the region's resident elephants were particularly aggressive in dealing with these newcomers. As a result, the newcomers had no choice but to match that aggression.

Hence, I only knew Chilonis as an aggressive matriarch, which is why she had been named after a Spartan princess and queen. She had become even more aggressive during this particularly dry year. Under normal circumstances, a thirsty family barreling in for a drink would not be screaming and roaring like this, but claiming this vital resource had become imperative for every family group—all of which had more babies than usual this year, due to better rains in the previous two years during gestation. Perhaps Chilonis's new baby was the reason for the intensity of her vocalizations, which caught my attention. We had named her baby Cleo and were looking forward to watching how the calf of such an impressive matriarch would make his way in the world.

Only twenty minutes earlier, Ursula and the Goddesses had made their own stand for the watering hole, forcibly removing a low-ranking family we called the Lost Girls. Ursula was formidable, but Venus de Milo, the "general" of the Goddesses, tended to do all the dirty work. Venus had hurled herself like a kamikaze fighter pilot on a suicide mission at any elephants that stood in her family's way. When the Lost Girl family didn't yield, Venus came to blows with the matriarch, Tiger Lily, in the same way that male elephant titans fight to the death for a mate. Only, in this case, the battle was over water rights.

Little did the Goddesses know that Chilonis had been in the background, waiting to make her move. The Scimitars were higher ranking than the Goddesses, in the same way the Goddesses dominated the Lost Girls, and when Chilonis's battle cry was heard, the Goddesses didn't hesitate. Against such a powerful adversary, it was not worth the risk of coming to blows again. Withdrawing the troops was the smartest option.

Ursula, Venus de Milo, and their family cut the mud bath short and made a fast retreat out of the pan and into the forest to the south. There, they engaged in a dust bath, all the while hoping for another turn at the waterhole before leaving the area.

THE BATTLE CRY is just one of many ritualized vocalizations that serve a very specific purpose—to announce an all-out war or dispute over territory. Many animals have such a cry, one that's sounded before, during, or after vanquishing the enemy—real or perceived. The battle cry or call to arms serves the very important purpose of galvanizing a group prior to engaging in an attack. This call might also help to avoid physical aggression if the opponent retreats.

Such rituals are dramatized in war movies, where a general typically delivers a rallying speech in front of a group of soldiers. The soldiers then respond with deafening roars or chants that embolden the troops. Chimpanzees engage in a similar vocal ritual with screams and hoots after slaying a neighboring male with whom they have been at war.

Engaging in these chants and aggressive vocalizations releases endorphins that serve to strengthen bonds within the group. This kind of vocal ritual has been used throughout time to express power and confidence during battle. For the same reason, shouts, chants, and rallying cries are an integral part of modern professional sports. And it's not just the players but sports fans themselves who are impacted by their own cheering and chanting. The roar of the crowd is quite literally a shared experience, one in which the spectators share and mirror the same physiological and psychological reactions. When their team is doing well, fans experience surges of dopamine and adrenaline, and increased heart rates and blood pressure, as if they were the ones playing. In fact, Canadian doctors caution hockey fans to not take hockey games so seriously or they might give themselves a heart attack.

Experiencing similar physiological states as the athlete playing the game is attributable to what are called "mirror neurons" in the brain. When we see someone else doing or experiencing something, we mirror the emotions connected to the action inside ourselves. When our team loses, we feel the loss, yell in frustration, perhaps scream in anguish and disappointment, and we might even become depressed for days. Does this mean we shouldn't watch sports or care who wins?

No, but it means that we should recognize the power of our vocalizations and how they affect us and those around us.

Fighting and competition aren't the only inspirations for vocal rituals. Ritual shouts can express intense joy or relief after a hard-won victory. The Greek ritual of breaking plates and yelling "*Opa!*" is a very positive release of emotions that is thought to ward off evil and may have its origins in grieving. Vocalizations are used to initiate action, such as when elephants give "let's go" rumbles or when gorillas increasingly grunt before a group departs an area. Vocal rituals can delineate boundaries or claim a territory, like lion roars, wolf howls, or howler monkey calls at dawn and dusk. They are also an important aspect of many courting rituals such as the elk bugle, and they are used by the red deer to assess the quality of a mate, and by the red squirrel as a tool to find kin.

Spoken rituals influence our emotions. They are a strong bonding mechanism between loved ones, friends, and associates. For instance, how do we usually calm a crying baby? By repeating sweet nothings in a soft voice or soothing them with a lullaby till they drift off to sleep.

Most mammalian vocalizations are produced from rhythmic patterns of muscle contractions within the larynx, the organ considered our "voice box," which generates pressure waves that resonate within the vocal chords. The larynx produces many different types of sounds like roaring, bellowing, barking, grunting, singing, chanting, and speaking.

Sounds are vibrations, which can be "heard" through air, water, and the ground. Elephants might detect certain vocalizations as vibrations in the ground, while many animals communicate, not by voice, but by vibrating material like a leaf or grass stalk. Both human and nonhuman animals can detect vibrations through a bone-conduction pathway, from the skull to the middle or inner ear, or via the sense of touch through the somatosensory, or vibration, pathway. Some animals, such as the elephant and the golden mole, have specialized ears that facilitate bone conduction—the elephant being several orders of magnitude more sensitive to bone conduction than humans.

In other words, animals don't need a spoken language to communicate very effectively using sound. Vocal rituals in modern, social vertebrates can be traced back four hundred million years. That said, the evolution of human language might be due in part to the FoxP2 gene, which many animals have, and which is not necessarily confined to language-associated areas of the brain. Broadly, the FoxP2 gene is expressed in relation to vocalizations in songbirds and mice, as well as other types of social communication in fish, frogs, and others. In the case of the zebra finch, it is expressed with increasing intensity during song-learning, when young birds memorize melodies sung by their elders.

For humans, spoken language has long been a necessary element of social cooperation and even of our survival as a species. It is so important, in fact, that it is curious that language did not evolve in our nearest relatives. Yet other species, like the chimpanzee and the African wild dog, for example, can still communicate very successfully without verbal language. Like us, they can successfully accomplish complex tasks, form coalitions, engage in cooperative hunting, and so on.

Some scientists attribute the evolution of language to the evolution of more sophisticated tools. As early humans created more advanced tools, they needed some way to explain to others the intricate steps required to make these tools. Language may have emerged to convey these steps, and therefore, it followed a similar pattern of rules and ordering. This laid the groundwork for a formalized ritual of communication through words.

VOCALIZATIONS AFFECT BOTH sender and receiver. All social animals are impacted physiologically when they create and hear vocalizations. This physiological response is deeply influenced by *the intent* *spoken rituals*

behind the vocalization. If the sender has positive intentions—say, a man serenading a lover in courtship—the hormone oxytocin gets released in both the singer and the listener.

Oxytocin is nicknamed the "bonding hormone," since it induces feelings of trust and connection. This is one reason lullabies are so effective with babies at bedtime.

For tropical birds like the swamp boubou, males and females sing in a remarkably coordinated duet ritual, where one calls right after the other in a continuous sequence. Or the two call at the same time. The male's call is tonal and sounds like a bell, whereas the female's call is harsh and choppy sounding. This duet is thought to facilitate pair-bonding. It is also used as a way for each bird to keep track of the other within dense foliage and to define their territory together.

Male serenades can have a broader impact than just bonding alone. Researchers have found that male vocalizations prime female hormones for reproduction. When repeated vocalizations of the male ring dove are played to a female, even without the male present, it causes her ovarian follicles to double in size. Over time, researchers realized that the female vocalizations by themselves turn out to have the greatest impact of all. This phenomenon of vocal stimulation of follicles is similar to visual displays stimulating a higher birth rate in houbara bustards and has been documented in other bird species such as the budgerigars, canaries, and the little blue penguin. Many animals serenade for very practical reasons, whether to another or to oneself, even if they can't play Spanish guitar.

Not only can vocalizations influence the physiology of the listener, they can reveal the psychological state of the caller. For example, when domestic pigs are separated from their companions, their adrenaline levels go up, and this increases their "squeal-grunt" vocalization. Adrenaline, the fight-or-flight hormone, gets released under stress, excitement, or threat of danger. For pigs, grunting is a more relaxed vocalization, but this diminishes under the duress of separation. This is how psychological states can be conveyed and understood through vocalizations, even without formal language.

Humans listen this way all the time, evaluating tone and inflection to understand someone's psychological state. People don't always say what they really feel—and sometimes we can't express our emotions clearly—but we learn what's unsaid just by listening to the *way* people speak.

The pitch of vocalizations conveys much information. For instance, low-frequency sounds often correlate with larger body size, or perceived body size. By projecting a lower-frequency call, the signaler provides information about their size. A larger individual could intimidate or discourage a potential attacker by using a deep-sounding voice. Or a low-pitched male call can serve to attract a mate, since females of many species often assess health through the body size of a potential partner.

Among social animals, most species send "honest" information about themselves in their vocalized messages. A wolf emitting a bark-howl is sending the message that it will defend the pack and territory, and others treat this seriously. Similarly, a male lion entering an area will roar its intention to kill any cubs that are not his own. When females hear this roar, they immediately hide their cubs from this murderous intruder, who is intent on starting a new pride with only his offspring to feed.

On the other hand, some frogs cheat. Males might use an amplifier like a drainpipe to simulate a lower-frequency call. This signals that the frog is large, and thus a more attractive mate—even if he is not.

People cheat, too, or at least, they try to influence impressions through tone of voice. A speaker at a press conference, professional presentation, or board meeting may use a lower voice to give off a sense of confidence and authority. This tone is more likely to win over an audience.

The loudness of a vocal ritual can also hold significant information. During the ritual roaring of red deer in mating season, a female red deer can decide on her mate by assessing the volume of the male roars. Testosterone influences the laryngeal muscles of the deer, making it possible to roar louder, so the female red deer knows that the louder the call, the higher the testosterone levels of the caller.

This is also true of ritual territorial calls. The wahoo calls of the male baboon, for example, allow other males to assess the dangers of a potential encounter based on the characteristics of the caller's vocalizations. In the male common loon, a lengthening of his yodel vocalization is an announcement of his threat to take over a territory. For the red squirrel, territorial vocalizations provide the opportunity to discriminate kin from nonkin. This recognition eases tensions among relatives, who don't have to worry about aggressive encounters with one another.

The territorial "great call" of the female white-handed gibbon is a very loud vocalization produced at one hundred decibels, and it provides information about her age and health status. This haunting call serves as an honest signal with regard to her health and the quality of the resources she is defending. It also helps her attract a mate.

Throughout the natural world, vocal rituals make it possible for mothers and offspring to identify and find each other in a crowd, whether in a crowded shopping mall or a crowded bird or seal colony. This is critical at a distance, when olfactory recognition (which is also used) isn't possible. Studies have shown that a human baby learns to recognize Mom's voice from inside the womb. One reason hearing Mom is so calming could be because babies know the person most likely to take care of them is near.

Coordinated vocal rituals evolved for predators to keep track of one another during a hunt. For example, hyenas, coyotes, wolves, chimpanzees, and other predators use vocal cues in order to distinguish the position of each group member in the dark and over the space of sometimes miles. An elephant's low-frequency rumble vocalization can travel several miles through the air and helps everyone keep tabs on other family members while foraging.

IF YOU REALLY want sound to be heard at a distance, you need amplification, and humans aren't the only species to figure this out. In the ocean, whales communicate very long distances by taking advantage of the SOFAR (sound fixing and ranging) channel deep in the ocean. This channel creates a boundary above and below a horizontal layer that conserves the signal's vibrational energy and allows sounds to travel thousands of miles.

On land, a temperature inversion at dawn and dusk creates physical properties in the air that are similar to the ocean's SOFAR channel, and this also allows for long-distance transmissions. At nightfall, temperatures get cooler close to the ground, even as a warmer temperature layer sits above the cold layer. When dense clouds form above this warmer layer, sounds bounce off of and get sandwiched into the warmer layer above the colder ground, and these sounds travel faster and farther than usual.

This is why so many animals communicate at dawn and dusk, during the brief window when the sun rises and sets and sound transmission is at its peak. However, sounds emitted in the air are still limited by how far they can travel—six miles or so is the outer limit under most conditions. In the ground, there is no limit to how far a vibration can travel. This is why the vibrations from an elephant's low-frequency rumble might be detected much farther through the ground than through the air.

Long before modern humans invented the telegraph and the telephone, people discovered how to send vocal, vibrational, and musical signals long distances. The first known long-distance instrument was the bullroarer, which dates back to the Paleolithic some eighteen thousand years ago. The bullroarer consists of a rectangular slat of wood attached to a cord. The cord is twisted, and when swung in a large circle, the wood at the end makes a sound. This instrument was used to call up spirits and to communicate long distances. It has been found in Europe, Asia, Africa, America, and Australia and is still used in rituals worldwide.

Native Americans used song and drumming rituals, along with smoke signals, to communicate with neighboring tribes, such as to ask for assistance or announce an illness. The traditional talking drums of West Africa were used for a similar purpose. The pitch of the hourglass-shaped drum, which has two drumheads, can be varied to replicate the tonal patterns of speech, allowing for more sophisticated messages that can be transmitted long distances.

Because African languages are tonal in nature, similar to Chinese, complex messaging could be encoded in the drumbeats. This meant the drums could generate phrases that were learned and passed on. Drumming could describe phases of the moon or a footpath to take to hide from or ambush the enemy. Lower frequencies indicated a man, and higher frequencies a woman.

Other cultures used wind instruments. About a thousand years ago, the Aboriginals in Australia invented the didgeridoo, which generates low-frequency sounds, and it has been used in cultural ceremonies ever since. The length of the instrument determines the pitch; the lower the pitch, the farther the signal can travel.

Yodeling is a vocal ritual with a long history that is still a major feature of folk music from Switzerland. It began as a method for sheep herders in the Swiss Alps to call in their flocks from far distances and to communicate with fellow herders. In yodeling, consonants are used to launch tones from low to high, giving the yodel its unique and far-reaching quality.

Although yodeling is most commonly associated with the Swiss, many other cultures have their own versions. In the first account of yodeling, in 397 CE, the Roman emperor complained about the noise of herders communicating across valleys. There is speculation that the practice evolved when animals were first domesticated over ten thousand years ago. West African Pygmies have a yodel ritual to enhance their intimate relationship with forest spirits before going on a big game hunt, or *maka*, to celebrate cultural rites such as masculinity and ethnic identity. They yodel to ward off danger and evil spirits and to ensure—and later celebrate—a successful hunt.

When I first hiked the trail to the Matterhorn outside of Zermatt, I was blown away by the beauty and history of the Alps. I developed a deep appreciation for yodeling as a ritual and how it can communicate messages long distances between mountain peaks. Almost every culture and musical genre—from classical, rock, R&B, jazz, and country to opera—uses yodeling. The song "The Lion Sleeps Tonight" incorporates yodeling that you may not realize is there. Even Adele uses a form of yodeling in her songs.

Yodeling is so deeply rooted in the psychology of the Swiss that, in the seventeenth century, the practice was banned in the presence of Swiss mercenary soldiers. Apparently, upon hearing these alpine songs, too many of them suffered from homesickness, went AWOL, or even died.

OF COURSE, WE DON'T NEED to yodel to communicate between mountains anymore. Using digital technology, we can speak and be heard down the block, across the country, and around the world in an instant. But is our ability to share information on the internet and to talk via text and satellite the same as speaking to someone in person? For all the revolutionary benefits of digital technology, face-to-face dialogue is still important and provides unique benefits that we shouldn't take for granted.

spoken rituals

Our bodies benefit from vocal exchanges. For example, having a conversation to catch up with your loved ones at the end of the day is healthier than eating dinner in front of the television to decompress. When I was growing up, my parents made a conscious decision not to have television at our dinner table, despite my older brother's many objections. Even though it was hard for my parents to pry quality conversation out of their four teenagers, I think we were better off without the TV.

Admittedly, not all the conversation was uplifting. There were awkward silences and strained arguments when my brother refused to finish his vegetables before dessert. However, we look back on those moments and laugh, and sharing our memories and experiences of our family dinner table brings us closer together today. Our memories may not have been as fond if we'd been watching TV. However, a shared experience is better than none at all, the point being that a shared spoken experience is powerful and important.

Studies show that spoken rituals increase our level of commitment, such as when we speak wedding vows or when graduating medical students recite the Hippocratic oath. Verbally expressing how we feel, speaking our thoughts out loud, helps us process our feelings, especially difficult emotions of grief, jealousy, and frustration. The simple act of sharing our thoughts with another person or a group reduces stress and improves our well-being; it creates more fulfilling relationships. This is why individual and group therapy is so restorative. Beyond whatever we actually say, speaking to someone in person and being heard is a healing experience.

The same is the case for simply expressing ourselves or singing out loud. Think about how good it feels to sing in the shower or to sing along to your favorite song at a concert. In a study of music therapy, researchers found that singing has an even greater impact on individual psychological well-being than talking does. Singing in a group lowers cortisol levels and increases the expression of oxytocin, creating a sense of connection and relaxation within the group. Singing also reduces depression and feelings of loneliness.

The elation we feel when singing—in the car, in a choir, in a karaoke bar with strangers—comes from endorphins, responsible for our feelings of pleasure. As it turns out, endorphins are released whether we are a good singer or not.

Singing is now being used to manage the psychological side effects of a range of disorders, including arthritis, lung problems, chronic pain, and cancer. Studies have shown that music therapy can reduce anxiety and improve quality of life, while sometimes even reducing symptoms and side effects of disease.

Despite all these health benefits, many scientists believe music and song are not evolutionary adaptive behaviors, that music has no survival value. Their opinion is that music evolved from a higher order intelligence, purely for aesthetic reasons, or perhaps solely for pleasure. Some aesthetic philosophers believe that the defining characteristic of the arts is that they serve no practical purpose other than pleasure, and that pleasure isn't essential to survival.

What makes more sense to me is what cognitive scientist David Huron proposes. He suggests that pleasure-seeking behaviors can also be evolutionary adaptive behaviors, and he notes how our most pleasurable activities—like eating and sex—are clearly linked to survival. Besides, evidence of human music-making dates back to the Paleolithic era.

The earliest known musical instrument, found in Slovenia, is estimated to be between forty-three and eighty-two thousand years old. It is a bone flute carved from the femur of the now-extinct European bear. Huron reviews eight theories for how music may have evolved and why it has been important for human survival. First, it helped attract a mate. Then and now, music facilitates social cohesion and group coordination, it reduces conflict, and it benefits perceptual development, hearing, and motor skills. Further, as early humans became more efficient at gathering food and had more time on their hands at night, engaging in music would have been a safe and enjoyable way to pass the time.

In addition to these life-saving perks, music has been used to pass along information from generation to generation, such as the "song lines" of Aboriginal Australians. This group has the longest continuous cultural history on Earth, due to its passing down of stories about social law and practical information about plants. Knowledge of animal behaviors used for successful hunting was conveyed through song, dance, art, and music.

Music, and music-makers, have evolved across many different cultures over centuries. Given this history, and how music persists across generations, it's hard to see how the art of song is not adaptive and part of human survival. Music-making may in fact be in our genes.

Not only singing but listening to music is a pleasurable activity, releasing dopamine, that has positive physiological effects and health benefits. Listening to music arouses strong emotions, which is pleasurable and rewarding in itself, regardless of any physical or practical benefit.

Finally, although many benefits of music therapy have been quantified, such as the feeling of calm that patients with Alzheimer's disease experience from an increase in melatonin levels, some benefits of music are harder to measure. All religions use music, singing, and chanting to generate a sense of awe and humility as well as to engender a deeper sense of community. When singing hymns or chanting the Sanskrit word *om* during meditation, we use ourselves as musical instruments, and the vibrations generated within are very therapeutic to both mind and body.

A FEW DAYS after I witnessed Chilonis's battle cry at the waterhole in Etosha National Park, an elephantine family argument was brewing of a different kind. Late one afternoon, the Goddesses were leaving the waterhole and signaling their departure with a turn-taking sequence of "let's go" rumbles. From the patterns of the exchange, I could tell they were disagreeing about the direction the family should take after their evening drink.

Mona Lisa, another high-ranking female within the family, had convinced a portion of the family to head northwest along the elephant path we fondly referred to as "Bolus Boulevard." At the same time, Ursula, the matriarch, and her daughter, Slit Ear, were leading their contingent out to the south, rumbling as they went. Now both subfamilies stood at opposite edges of the clearing in a vocal tug-of-war—each seeming to communicate that their decision was the right one.

Since they were facing in opposite directions and spaced at least five hundred feet apart, a naïve observer wouldn't think they were communicating to one another at all. They looked like two family groups making their separate ways toward two different destinations. But I recognized disagreement from the loudness of the calls and the rise in pitch.

Mona Lisa rumbled and then Ursula and Slit Ear rumbled in a growling retort, one after the other. Even during a disagreement, there was an element of turn-taking to their discourse. Eventually, Mona Lisa won. Ursula and Slit Ear changed course and followed Mona Lisa's path to the northwest. This was a surprising turn of events, given that Ursula was the matriarch; it went without saying that everyone should have fallen in line and followed her lead. Yet the elephantine world, like the human world, does not play out in black and white. Decisions of the leader don't always go unchallenged.

However, I was struck by the coordinated nature of these departure vocalizations, even in the midst of a dispute. Each of the mature females was given her fair share of time to voice her opinion. Waiting for each caller to finish before another vocalized also served to extend the length of their vocal signals, so that they would be more easily heard in the distance. If another group were eavesdropping during this departure, they would know that the waterhole was now available. Paying attention to these vocal cues could help to avoid conflict over access to the prized water resource.

In contrast, our own vocal disputes are full of overlapping discord. In the most heated situations, turn-taking is often the first thing we forgo. We could learn a thing or two from the measured exchanges between female elephants.

This happens even during normal conversations. How many times have we come away from lunch with an old friend feeling that we might have dominated the airspace and didn't let our friend get a word in edgewise? Or the opposite, that our old friend sucked up all the oxygen in the room and we left feeling dissatisfied? Having an even exchange requires practice and mindfulness. Since conversation is critical to well-being, turn-taking is an important skill to keep in mind during your next phone or dinner conversation. Remember the elephants and wait until another has finished speaking before chiming in.

WHILE PASSIVELY LISTENING to sounds at my elephant field site, I feel grounded. Jackals call around the perimeter of the clearing at sunset with a chorus that speaks to the soul. Soon, elephants begin to fill the night air with roars, bellows, trumpets, and deep low rumbles. At some point, in the middle of the night, a hyena scales an octave with its contact call, searching for its pack. I may not talk much during these times, because listening to the evening's chorus feels restorative. These bouts of silence can become somewhat addictive at times, and as camp leader, I know I can't allow myself to dwell for too long.

A full moon can be the busiest time for animal vocalizations. Black rhinos are particularly vocal at this time. Territorial males huff, puff, and bellow, staking their claims over the water just as the elephants do. Elephant families arrive at the watering hole around sunset and throughout much of the night; the roaring, rumbling, and screaming of pachyderms fills the air.

At the end of the season, a persistent wind picks up, causing families to hold off visiting during the day, even though the need for water is dire. After the full moon, elephant families arrive later and later each night leading up to the new moon.

I can see a number of families gathering on the distant tree line, beyond the clearing, waiting for the wind to abate. Around five o'clock, they come barreling in. The onslaught of families erupting from the trees is like a scene out of *Jurassic Park*—each group jockeying for the best access to water. The rest of the season's regular families typically visit later in the night, their quiet rumblings in the darkness often the only hint of their presence. The pace finally slows to a halt well after midnight. The intermittent padding of a single bull coming in for a drink serves as the quiet heartbeat of Mushara until dawn, when the lion's territorial roars break the silence.

Each year, as the end of the field season draws near, the thought of leaving this extraordinary place weighs heavily on my mind. Efforts are torn between the necessary packing up and the longing to stay at this primal oasis. The sunsets glow red on the pan, while thirsty elephants line the edge of the spring. Giraffe necks define the horizon.

Somewhere in the distance, a jackal comes upon a lion and barks in warning, setting off a jackal night chorus. As I put my head on my pillow, the lions roar. The pressure waves created by their powerful vocal ritual hit my chest, turning on my fear response, reminding me that I, too, am prey in this environment.

As for Chilonis and her battle cry, I finally understood the motivation behind her war-like behavior. I had been watching her carefully for the remainder of the days before we left, and on the second to last day of the season, Chilonis arrived with her family but with no calf underfoot. I looked everywhere to make sure that

Cleo wasn't hiding behind another family member's legs. He was nowhere to be found.

Chilonis's calf was gone. I don't know how it happened, but looking back, it appears that Chilonis understood the circumstances were not good for her calf, and she was all the more aggressive in defending her water rights with her new baby's fitness to protect. The family wariness of their matriarch's intent, and the fragile environmental situation that her calf was born into, may have sparked the extra intensity of their roaring and bellowing when they chased Ursula and the Goddesses from the waterhole. For a mother, the stakes were as high as they could be.

THE GRAVITY OF PROXIMITY, POSTURES, AND EXPRESSION

"He who knows, does not speak. He who speaks, does not know."
—LAO TZU

WITHIN A THICK ASPEN AND lodgepole pine forest covered in a blanket of snow, two wolves catch sight of each other. When Lakota is approached by his brother, Kamots, of the Sawtooth wolf pack, he immediately hunches his shoulders, crouches down, and holds his head low in supplication. He wants to make himself appear as nonthreatening as possible—despite being the largest wolf in the pack. This ritualized visual gesture is an important signal of submission in wolves.

Kamots towers over Lakota and bares his teeth, a visual reminder of his station. Lakota remains low and delicately sticks out his tongue. He gently licks at Kamots's intimidating snarl.

Lakota's visual display lets Kamots know that Lakota recognizes his brother's position at the top of the hierarchy. As such, he would not show any aggression toward him, regardless of his size.

Despite their deep bond, the unspoken ritual of supplicating postures between brothers was necessary body language in order to avoid being put in his place. In this wolf pack, Kamots was the alpha and Lakota was the omega.

Clearly, size is not the only thing that determines dominance in wolves, as Lakota was a bit submissive in character. And he was easily bullied after he fell into the omega position as a young adult.

This pair of wolf brothers were the study subjects of Jim and Jamie Dutcher. The goal of their six-year research project was to understand wolf society in Idaho's Sawtooth Mountains, and they followed the Sawtooth pack from pups to adulthood in the 1990s.

Dominance can fluctuate in wolf packs, depending on the very fluid social dynamics. This explains why status is reinforced frequently. Fortunately, in the Sawtooth wolf pack, Lakota had a buddy, Matsi, the second highest ranking male in the pack. Matsi sought out the omega to play with, and he often protected the omega from the bullying. The dynamic wasn't always perfect, however.

In a conversation with Jim and Jamie, they told me that one day Lakota and Matsi had a falling out. Lakota remained lying under a tree for a long while, not interacting with any of the others, until finally, Matsi approached. He stood over Lakota and peed all over him and that was it, transgression forgiven. Lakota was allowed to get up and go about his business.

Despite Lakota's display of nonthreatening postures, every now and again, the midranking wolves, Amani and Motomo, would still dominate him. This was performed as a means of extra insurance, to keep the linearity of the hierarchy clear. Amani and Motomo would also occasionally dominate Lakota if Lakota's howling grew too vocal when he joined in the nightly howling chorus.

It's upsetting to think that a wolf could ever be deprived of this primal ritual because of another's insecurities. As such, I was happy to learn that Lakota spent his later years with a different wolf occupying the omega position. And he apparently never picked on that wolf like others would have in his position in all his remaining years until he passed away in 2002.

Nature may seem incredibly cruel at times. While it's tough to be the lowest-ranking member of the pack, having a dominance hierarchy—or pecking order—within social groups has its advantages. These unspoken rituals within the Sawtooth wolf pack create a linear hierarchy that gives them important advantages when it comes to the coordinated activities of hunting or defending a territory. In fact, some of the most important conversations in the wild are held without words.

Unspoken rituals of dominance provide structure to societies. If individuals know their place, there are fewer life-threatening conflicts over resources. In cases where the dominant individual tends to be the leader, as is the case with an elephant matriarch, members of the group entrust their safety to her. In the case of the olive baboon, however, leadership is more democratic, whereby the highest-ranking individual is not necessarily the leader.

Just as unspoken rituals indicate someone's low ranking, others signal high rank. When the bull elephant in my study population, Smokey, is in musth, he appears on the horizon as if the Earth suddenly erupted and spewed out a giant granite god onto the savanna. The alpha male floats toward the tree line and enters the clearing as a larger-than-life elephant, full of pomp and circumstance, befitting his title as the true king of the jungle. Smokey announces his hormonal state of musth by his exaggerated gait—his back foot overstepping his front foot, head and shoulders held high as he swings his trunk from side to side. He waves one ear forward and then the other, all while curling his trunk across his face as he marches to the waterhole.

Like any animal society, much of the human world is made up of subtle and unspoken rituals that signal power, weakness, aggression, and invitation. Like other social animals, we, too, benefit from certain postures that fill us with confidence in a threatening or intimidating situation. Males in many species strut in an attempt to intimidate another male. This display of confidence allows them to take over a territory, gain access to a mate, or climb in rank within a hierarchy.

In the *Anolis* lizard, the males are constantly strutting about, defending or attempting to take over a territory. This happens on a daily basis in my own backyard. The male *Anolis* does a number of push-ups with his front legs, while flaring his colorful throat dewlap, in and out, in an attempt to intimidate another male. A lot of energy is spent on this behavior, and the species continues to persist, so it must be effective. It isn't just the anoles who throw their weight around. From lizards to humans to elephants, the unspoken ritual of posturing is a successful strategy for making a statement.

Posturing is usually coupled with a hormonal cue. It turns out that the pose that we strike while seated in an important business meeting can garner us an advantage with our colleagues. We can exude confidence by how far back we lean and, depending on the situation, even putting our feet up on our desk. We can reap the benefits of these power poses prior to a meeting, as a warm-up, similar to what an actor does to warm up before a performance. Studies have shown that a spike of testosterone accompanies power poses that further project confidence to influence others in a meeting. Additionally, striking such a power pose reduces stress.

If you think about it, when a male elephant in musth is strutting around, curling his trunk across his face, announcing his dominance, it's not so different from an overly aggressive human marching around the boardroom, swinging arms around to gesticulate dominance. Both of these behaviors correspond with higher testosterone levels in the signaler, whether male or female, as females also express testosterone. When elephants and orangutans exhibit these behaviors, there is a decrease in testosterone expression in the surrounding individuals.

The reverse behaviors also convey an important signal: Hunching forward, crossing legs, holding one's head down, and making minimal eye contact can convey insecurity. Practicing power poses should not be taken lightly if one is serious about nailing a job interview. This is also relevant for approaching a boss to get a raise or gain respect during a challenging negotiation. Or simply when trying to get a point across in a room full of outsized egos.

Paying attention to body language is important in any setting. Subtle gestures are just as impactful as overt gestures in ritualized signals. Their effect on others and ourselves is quantifiable. Body language and facial expressions—whether threatening, conciliatory, fearful, or humbling—all impact our physical and mental well-being. Both the signaler and receiver can be influenced by them. These gestures may even impact our reproductive health.

RITUALIZED OLFACTORY DISPLAYS are also important unspoken rituals in the context of dominance and mate-finding. For a male elephant in musth, his far-reaching olfactory cues serve three purposes—intimidating other males in the vicinity, suppressing others from going into musth, and attracting females that are in estrus. The next time you smell a little too much cologne in the boardroom or nightclub, you might wonder whether intimidation or attraction is the intention—or both.

In the case of Smokey, if there are any male elephants present that could be potential competitors, his urine dribbling (and therefore scent dispersal) accelerates as he nears the group. The same happens if he picks up on the scent trail of any other young musth bull who may have visited the watering hole recently.

For a decade, this was true for Smokey and the young bull Ozzie when both were in musth. Ozzie's movements were always at least one day ahead of Smokey's schedule, and not without considerable effort. He spent an enormous amount of time planning his departure trajectory. He'd place his trunk on the ground for long periods, presumably assessing the whereabouts of his nemesis. It seemed as if

Ozzie could detect Smokey's sovereign gait through vibrations emitted within the ground. Once he picked up on Smokey's location, he would choose to travel in the opposite direction.

In the days following Ozzie's visits, Smokey would typically arrive in a musth-filled rage, tracing Ozzie's trail of urine dribble from the previous day. He seemed livid at Ozzie's defiance of an age-old gentleman's agreement—that one should not enter the hormonal state of musth until at least the age of twenty-five. As male elephants enter musth serially, as opposed to all at the same time, the older musth bulls would have seniority during the period of time when most females were in estrus. Ozzie was clearly not paying attention to the rules, gentlemanly or otherwise.

Smokey's unspoken rituals kept all the other younger and less-dominant male elephants at bay. Although he was tolerant of older bulls in musth, there was something different about Ozzie. Even formidable bullies couldn't evict Ozzie from his throne. Having witnessed his aggression toward some of Mushara's most gentle, senior elephants, like Brendan and Abe, I, too, found myself wanting to see Ozzie put in his place. Smokey seemed like the only possible candidate to unseat the young bull who had defied nature for so long.

One afternoon, while Ozzie was exerting his dominance and clearing the area of all the other males, it finally happened. Smokey entered the clearing in his usual way—in full musth swag, curling his trunk over his head like a lasso, waving his ears back and forth, marching his way to the waterhole while dribbling urine.

This was the first time that I had ever seen Ozzie and Smokey occupy the same place at the same time, after having watched Smokey chase after the odor plume of Ozzie's lingering testosterone for all those years. In a lapse of vigilance, Ozzie's carefully orchestrated movements to avoid Smokey had failed, and I anticipated a serious clash. Ozzie's decade-long power trip was about to come to an abrupt and violent end.

unspoken rituals

Ozzie caught sight of Smokey and remained where he was, standing in the pan, carefully observing the approach of his nemesis. He did not retreat nor show any sign of either aggression or submission. Smokey held his head up high and ears out wide. Instead of backing down as Smokey neared, Ozzie held his ears out in defiance.

The older and more dominant musth bull, Smokey, towered over Ozzie, and should have been intimidating. Yet the meeting didn't erupt into aggression, much less all-out combat. Given Smokey's reaction to picking up Ozzie's scent trail over the years, I expected a little more fanfare.

The moment came and went as these two rivals took stock of each other in the flesh. To the untrained eye, it appeared as if nothing of importance transpired. To the trained eye, Ozzie was indeed intimidated. Moments after his initial show of defiance, Ozzie held his ears flat to his side and relaxed his posture as Smokey walked up to him calmly but with purpose. The jig was up. No contest.

Even more surprising, Ozzie positioned his backside to allow Smokey to inspect his penis in submission—an incredibly risky maneuver, given that Smokey could have delivered a lethal blow with his tusks in any number of ways. This act signaled Ozzie's acceptance of Smokey's dominance. No elaborate musth display from Smokey was necessary.

After Smokey inspected Ozzie, Ozzie slunk away with low shoulders and kept his distance, as if trying to draw as little attention to himself as possible. This was certainly not Ozzie's default behavior in the presence of other older bulls, in musth or otherwise.

Two days went by, and then Ozzie reappeared at the watering hole, no longer in musth. I barely recognized Ozzie without his previous swagger and extreme aggression. Without a fuss, Ozzie quietly slipped into the clearing with another young bull, Kelly. The only remnant of musth was a slight, temporal staining on his cheeks.

A simple raised head and stern look from the majestic Smokey was all it took to start a process of hormonal suppression that, within two days, resulted in Ozzie's submission. Ozzie's careful calculations of Smokey's whereabouts appear to have been critical to his reign of terror. Nature had finally made a course correction. As one of my volunteers said of Smokey, "Ozzie wasn't worth spilling his Scotch over."

It turns out, even in the elephant world, no one likes a tyrant. Character is an important determining feature of dominance. This is one of the many interesting things that I've learned in my work studying male elephant society. Smokey never threw his weight around, despite the many opportunities that he had to do so. He was much admired by both male and female elephants.

Similarly, the most dominant bull in my study, Greg, had a fascinating balance between carrot and stick—knowing just when to dole out the attentions of a full body rub or to exact punishment with a tusk jab. He reeled in the youngsters by allowing them to cuddle up to him and even suck on his tusk. Whereas he came down hard on the bullies. Despite the intolerance he showed for the mid-to-low-ranking bullies, he still let them be part of the group, and they seemed eager to stay.

Greg was the social glue that held his cohort of fifteen individuals together for the six years that we studied his reign. He did this mostly through unspoken rituals. Eventually, a trunk wound left him permanently disabled. A large hole in his nostril caused half the water that he drank to spill out. Although he recovered his physical health, and returned to the top of the hierarchy the following year, he never returned to Mushara after that.

Today, years later, I have to assume the worst. One thing that makes me smile, despite feeling the constant loss of a great figurehead, is that the low-ranking bullies have now softened their approach to be more like Greg in their attempt to climb in rank.

Watching Ozzie lose his musth mojo was the perfect demonstration of why the presence of an older musth male is an important regulatory force within the population. Smokey's ability to hormonally suppress a poorly behaved subordinate highlighted the influence that hormones can have within any population.

There have been several cases of elephant introductions into South African parks that did not have elephants previously. When young males were introduced to the parks without the presence of older males, they went into the hormonal state of musth earlier than they would have in a natural population. They also exhibited deviant, aggressive behaviors toward other species, particularly rhinos.

Older male elephants were then introduced to the parks in the hope of curbing these aberrant behaviors. The mere presence of the older bulls caused the young males to go out of musth, and they became far less aggressive.

I found this to be true in my own studies of the role of aggression and hormones in male dominance hierarchies in wild elephants. Young males were more aggressive in years with heavy rains, since they had many different places to drink. They could avoid proximity with older males, causing their testosterone levels to elevate. Yet young males were less aggressive in dry years when they had to be in very close proximity with elders in order to access the few water resources that were available.

Our hormonal state is constantly being influenced by those around us, and it happens almost entirely without our notice. Take, for example, how quickly a group of women, living in a college dorm, automatically synchronize their monthly hormonal cycles, purely due to proximity.

Synchronized cycles, and thus synchronized births, are very common in social animals, particularly those with a high parental investment. The group benefits from crèches of young who are protected by one or a few. This allows other parents to gather food, as can be seen in penguins, flamingos, and other colonial populations. Synchronized births mean that antelope and other mammals don't have to forage for their offspring's food and benefit from the protection of the entire group.

Predators such as lions benefit from synchronized births by taking advantage of the communal rearing of cubs. For early nomadic humans, synchronized births would have benefited groups. Giving birth and dealing with fragile newborns could occur within the same window of time, thus minimizing the impact on a group that needed to migrate based on prey availability and seasonality.

We're not accustomed to thinking of another person's hormones having an influence on us, but we understand the influence of role models. We know that children need role models, and that positive parental figures or mentors are critical in order for teens to develop healthy goals in life. The importance of proximity can't be understated. Role models need to be seen, experienced, and interacted with in order to be most effective, whether it be at home or at school, within sports teams or extended families. This is why rituals that provide a bridge from adolescence to adulthood are so important. The bar and bat mitzvahs in Jewish culture are examples of such rites of passage.

UNSPOKEN RITUALS DON'T just signal location in a dominance hierarchy. Facial expressions are very powerful signals that can be important for coordinating social interaction, facilitating group cohesion, and maintaining individual social relationships.

Who would have thought that a simple flash of our pearly whites is an ancient, unspoken ritual? It's so important that it has been passed down through evolutionary time and is present in many of our distant, primate cousins. For a long time, researchers thought that what looked like a smile for the chimpanzee was really an expression of fear. A recent study shows that chimpanzees do, indeed, smile and laugh, in the same context that we do. For example, a baby chimpanzee smiles and laughs just like a human baby who is being tickled. Scientists believe that the smile ritual may have originated in a common ancestor and serves to facilitate social cohesion and strengthen bonds as well as to reassure and appease. Recent studies have shown that a smile is good for your health—and so is laughter. Both behaviors are contagious.

Gaze is another important unspoken ritual. Both parent and child get all warm and fuzzy when they look deeply into each other's eyes, triggering the release of oxytocin in the brain. The same thing happens between loved ones or even between a dog and its owner.

Gaze serves as an important "tell" of someone's mental state. The ability to infer someone else's mental state is referred to as "theory of mind." Humans develop this ability as toddlers, and researchers have demonstrated that gorillas have this ability too. Gorillas are able to ascertain intention based on a combination of body positioning and gaze. Theory of mind has also been demonstrated in chimpanzees, bonobos, and orangutans.

Gaze and gestures are such important signals that they have been formalized into a powerful form of communication called American Sign Language. Sign language is not just used with those who have hearing impairments. It's also used with babies before they learn to speak, and it's been used to communicate between species in a very compelling way.

While working on my book *Bridge to the Wild*, I had the good fortune of meeting the orangutans and gorillas at Zoo Atlanta. Some of these apes had learned to sign while part of several intelligence studies at the Yerkes National Primate Research Center in Atlanta. I was immediately struck by their ability to communicate basic needs, like asking for a blanket. From there, I quickly learned that cross-species communication (human and orangutan) went much deeper than this, including conveying emotion through sign.

I accompanied one of the veterinarians on morning rounds, and one of our stops was to visit a Borneo/Sumatran hybrid orangutan named Chantek, who was the first orangutan to learn sign language. He was thirty-six years old at the time and had been raised by anthropologist Lyn Miles at the University of Tennessee, Chattanooga.

Chantek was shy and, using sign, asked the keeper for a blanket. He tucked himself in, all the while watching me carefully as I asked questions about his history.

I discovered that Chantek had been raised as a human, with the hope that teaching him to express himself as a human might make it easier to understand his cognitive capacity. Over the course of nine years, he apparently developed the language skills of a seven- to eight-year-old child. When he lived on the campus of the University of Tennessee, Chattanooga, he was famous for giving driving directions to the local Dairy Queen, as he loved ice cream, though his all-time favorite treat was a hamburger.

After many years spent with his longtime caregiver and sign teacher, Lyn, Chantek got too large. The university administration decided they could not house Chantek safely, and he was returned to Yerkes, where he was born. There, he was kept in a very small cell, which was an extremely difficult adjustment. Having been raised with many freedoms and having developed close relationships with humans, such confinement seemed punitive.

Later, Lyn told me how difficult the transition was for both of them. She lamented that there was no halfway house for great apes raised as humans. Without a long-term plan, when they got too dangerous to manage in open facilities, it often didn't end well for these primates.

When Lyn visited Chantek in his new environment, he was very confused and distraught. She asked him how he was doing, and he signed that he was hurt. She asked him where he was hurt, and he signed, "feelings." He conveyed the desire to escape and go home to

unspoken rituals

Chattanooga with her. When she tried to explain that the situation was complicated and that she couldn't do that, he asked her, in sign, to open the door in secret so that no one would know she allowed him to escape.

Chantek had learned how to communicate his emotions using sign. He was extremely frustrated to be in a place where he couldn't communicate or move as he pleased or be with the people he was attached to. He felt even more trapped and confused. Sadly, he was kept under these conditions for the next eleven years, during which time Lyn couldn't see him very often. He withdrew and become dangerously overweight.

When I heard this story, my heart broke. I was hugely relieved when Zoo Atlanta eventually agreed to take him. There, he had much better environmental and social conditions, and Lyn and his other longtime caregivers could visit more frequently. The staff bonded with him and were very attentive to his needs, with a sensitivity to his past upbringing as a human. He was even able to enjoy making jewelry and painting.

While Chantek was growing up, Lyn took him to the zoo so that he could see other orangutans. She told me that he referred to them as "orange dogs." He did not see himself as one of them. He considered himself part-human and called himself an "orangutan person," which is certainly understandable considering how he was raised.

Lyn explained that, once he befriended the other orangutans at Zoo Atlanta, Chantek recognized that he had a special ability to communicate with the keepers that the others didn't have. He used his skills to show other orangutans what needed to be done while receiving veterinary care, reassuring them. With such fluency and expression of emotions, there was very little difference between Chantek and his human caretakers, despite being separated by twelve to sixteen million years of evolution.

Many more apes have displayed their emotional intelligence through sign. In one particularly profound case, a gorilla named

Michael, who had been born in the wild and was raised by the Gorilla Foundation, was taught over six hundred signs by both researchers and Koko, the first gorilla to learn sign. According to researchers, Michael signed to convey memories of his mother being killed by poachers, using signs to express the concepts "squash meat gorilla," "mouth tooth," "cry sharp-noise loud," "bad think-trouble look-face" and "cut/neck lip (girl) hole." Watching Michael tell his horrific tale through sign on the Kokoflix YouTube channel is very humbling.

I was extremely moved to witness cross-species communication using sign language. It was clear that signing and gestures offer many opportunities for us to understand cognition and the emotional intelligence of our closest relatives, the other great apes, as well as to foster interspecies communication with other intelligent animals, such as elephants.

Having been around captive elephants for years, I knew they could be trained to understand a wide range of vocabulary and signals for management. Behaviors like lifting a foot to check for cracks in their foot pad or opening their mouth to check on a tooth abscess were all communicated verbally by their trainer. At one point, I worked with an elephant named Donna, at the Oakland Zoo, to determine her sensitivity to detecting vibrations through her feet, where her trainer had her touch targets when she felt a vibration. It seemed like an obvious extension to train elephants to understand and convey concepts through a form of sign language appropriately designed for elephants. So, after the vibration-discrimination experiments, I couldn't help wanting to ask questions about long-term memory and intelligence, and I started laying the foundation for a shared sign language.

Everyone is familiar with the phrase "an elephant never forgets." Plenty of anecdotes are told of elephants remembering both humans and other elephants whom they shared a bond with decades in the past. Elephants also remember migratory routes, the knowledge of which is apparently passed down from generation to generation. At the end of the dry season, elephants can migrate hundreds of miles toward rain to obtain fresh food and water. Along the way, they use the same paths again and again and pass the same favorite fruit trees. This knowledge of specific migratory routes and resources gets passed from the matriarch to all the adult females in the family. Exactly how elephants use long-term memory to remember these routes has not been determined. At the time I was working with Donna, no study had been done to demonstrate that elephants could recall something from the past and make a plan for the future.

To address this question, I wanted to show that Donna could imagine an item in her mind and remember it. I started by teaching her to repeatedly identify a picture of an object that matched the real thing, in this case, a banana. Then, Colleen Kinzley, Donna's trainer, presented her with a real banana while two other volunteers and I gave her a choice between three cards—two white laminated cards with no image and a third card containing a picture of a bright yellow banana.

It quickly became obvious that a visual challenge was more difficult than a tactile challenge, since an elephant tends to use its ears and nose first before cueing in visually. However, lots of banana treats later, Donna figured out the challenge and had no problem recognizing and selecting the picture of the banana as a representation of the real thing when presented with an actual banana.

We had a long road ahead of us, but this was a thrilling first step toward understanding how an elephant's long-term memory works by showing that an elephant can imagine an object in its mind as a referent of the real thing. This brought us closer to understanding how an elephant might make plans about an object that they can't see. It was also a first step toward another ultimate goal of understanding the inner elephant in Donna and the potential of a basic sign language. Having witnessed the poaching crisis in Africa firsthand, I can't help imagining that if humans could understand the inner lives of elephants, we might better understand our impact on their lives and, ultimately, on their well-being. Since we can use sign language so successfully with great apes to understand their emotions, why not with other socially intelligent animals like elephants?

In everyday life, signing between pets and pet owners would strengthen bonds, improve communication, and facilitate a deeper understanding of our furry companions. My dog, Frodo, understands many words and my body language. The words *walk*, *treat*, *beach*, and *dinner* all get him excited. The word *bath*, however, does not.

Frodo has taught me many of his own signals, like scratching the sliding-glass door with his paw when he needs to go outside. He shakes his head so that his ears clap to wake me up in the middle of the night and get me to tuck him back under his blanket. He has a high-pitched rapid bark to tell me that something is out of place—like the time when our toilet backed up into the bathtub and flooded the living room with sewage, a very timely communication. By developing a form of canine sign language, we would provide even more opportunities to confirm that dogs are extremely emotionally intelligent.

The fact that a border collie named Chaser learned the words for over a thousand objects is a good indication of a dog's ability to discriminate spoken language. We already know that they are masters of subtle body language. I recently learned about a deaf dog named Blue as well as a colleague's cat that are being trained to use a form of sign. So there is huge potential for a deeper level of interspecies communication in our very own homes.

MANY UNSPOKEN RITUALS are done subconsciously in the context of dating or mate selection. Social scientist Sandy Pentland has written several books about honest, subconscious signaling and how these signals are perceived by the receiver. By scoring verbal cues and body language using a metric he calls "social signal processing," Pentland has developed a formula for interpreting body language.

Pentland applied this formula to speed dating to determine the outcome of interactions. He describes a successful encounter as a "tightly choreographed dance," where arm motions are in sync, along with head-nodding. A failed encounter, one with a low likelihood of a future date, is out of sync and has many awkward pauses.

All behaviors in the study were captured using sensors worn on the body—a tool that he nicknamed the "jerkometer," since it captured how much of a jerk a subject was during the speed-date interaction. He considered the give-and-take of conversational turn-taking and gesturing, along with the timing, energy, and variability within the exchange. His results indicated four components of honest signals. These included how much influence one potential date appeared to have over the other.

The presence of mimicking behaviors between individuals was also an important signal, as were matching activity levels and consistency within the interaction—without awkward gaps. We understand positive, honest signals as rituals of connection. When we observe these unspoken rituals in someone else, it helps generate connection, which is strengthened when we repeat these behaviors. Over repeated meetings, this bonds two people further.

An additional, interesting element of the speed-dating experiment was that men only requested a second date with women whose body language signaled that they had an interest in them. These men were able to assess the subconscious cues of the women sitting across from them and recognized the honest signals. In the end, they could tell which women were receptive and which weren't.

Honest signals are sometimes very subtle. They can be nuanced and subconscious, such as touching the face or hair in a certain way or positioning oneself more openly, toward the other person, as opposed to maintaining a cool posture. Nodding while the other is talking conveys engaged listening, as compared with someone who sits still and doesn't gesture at all. A simple gaze or smile could signal either encouragement or discouragement.

Perhaps it goes without saying that dance is an important unspoken ritual of courtship and connection, but it's more than that. The vibration of dancing can have a profound impact on the human body. Ecstatic dance has diverse cultural underpinnings, where dancers are unencumbered by choreography and succumb to an internal bodily rhythm that leads to feelings of ecstasy. This dance ritual has been practiced throughout human history and is a component of certain religions around the world.

Ecstatic dance was prominent in the ancient practice of African shamanism, along with rhythmic drumming, to alter a state of mind during a spiritual ceremony. It was also practiced during the classical Greek era among followers of the god Dionysus. Ecstatic dance was incorporated into modern dance in the 1970s by Gabrielle Roth and has had a following in today's club cultures in the form of raves. The ritual allows participants to express and explore their minds, bodies, and emotions through dance. If a simple pose can have an influence over our hormonal expression, it's not surprising that such extreme vibrations and erratic movements of the body have at least a similar impact on our psychological state.

Tactile gestures such as grooming are another set of unspoken rituals that serve a deeper purpose. Whether involving primates removing parasites or a human mother brushing her child's hair, these actions release the bonding hormone oxytocin, which promotes feelings of trust. Further, skin-to-skin contact during cuddling signals the adrenal gland to stop producing the stress hormone cortisol, which boosts our immune response and makes us healthier. Plus, touch releases serotonin and dopamine, which improve mood and curb depression. Since cuddling reduces stress, lowers blood pressure, and helps you relax, it has the added benefit of promoting better sleep. Couples who cuddle more have stronger, healthier relationships. Cuddling is so critical for newborns that pediatric wards often offer volunteer cuddle programs for babies whose parents are absent or busy at work.

Between cuddling, gazing, singing, power poses, proximity, sign language, and body language, unspoken rituals play an extremely important role in the lives of all social species. The supplicating gestures that the lowest ranking wolf, Lakota, exhibited within his pack to avoid confrontation were important to keep the peace. Having an awareness of our body language can be critical to our success, both within our professional relationships and in our personal lives. Think of that as you're walking down the street with your shoulders and head held high, and perhaps smile at the people you pass. This unspoken ritual makes everyone's lives richer.

7 · PLAY RITUALS

CATCH A LION BY ITS TAIL

> "We don't stop playing because we grow old;
> we grow old because we stop playing."
> —GEORGE BERNARD SHAW

AWOKE IN THE CRISP RED DAWN of a Namibian desert winter to a soft thudding of footfalls in the sand below. I could hear the faintest murmurings coming from the far edge of the clearing. I lay in my open tent, which was on the top floor of our metal research tower and twenty feet above the waterhole, and tried to figure out what I was hearing. This perch had been home for the past two months, as it is every summer, as it has a commanding view of the waterhole out to the edge of the forest, about six hundred feet in all directions.

Sounds travels well in the still air of early morning. I heard more padding footsteps and a hard thump—followed by even more thudding and what sounded like a full-on chase, tackle, snarl, and romp. The soft calls got closer, short moans, almost a cooing sound. As I became fully awake, I realized these noises could only mean one thing at Mushara—that the resident lioness, Paidia, and her cubs were back at the waterhole and engaged in a rigorous play bout, frisky as ever.

As quietly as I could, I rolled over in my sleeping bag to grab the binoculars. Fortunately, the door to the tent was always opened so that I could observe these moments throughout the night by just lifting my head in silence. Despite an opened tent being that much colder, the noise of a zipper or Velcro travels surprisingly far, and I also liked to sleep with an unobstructed view of the Milky Way and the Southern Cross. There was no other place on Earth I could do so with so little light pollution.

The cold bit at my now exposed hands and nose as I lay on my stomach and assessed the situation below. In the early morning light, I could see that it was indeed Paidia and her five yearling cubs. It had been a week since we had seen them, so I was very glad to have them back.

Tim and I had been hoping to photograph this pride when there was enough light to get a decent picture. Being the cautious mother that she was, Paidia never let her cubs linger much after dawn—which made for some very blurry images.

On this morning, however, the lions gave us a gift by staying a little longer than usual. It was toward the end of the season, and they were feeling more comfortable with our presence at the waterhole. I pulled back my fleece hood and grabbed the heavy camera that I kept right next to me, for early-morning moments like these.

Three of the cubs were enjoying playing next to the pillbox bunker, a concrete dugout we used to take close-up photos, located sixty-five feet from the waterhole. Two cubs leaned up against the side of the cement structure, cozying into piles of dry elephant dung to make a nest. As these two took in the warmth of the rising sun, a third cub dangled an oversized paw down from the top of the bunker, swatting at their heads. Startled, the two siblings leapt away as if this strange paw had just fallen, inexplicably, from the sky.

Inspired by its littermates' reaction, the third cub took his plan a step further. A few moments after his siblings had settled back into

play rituals

their dung nest to continue sunning themselves, he stalked them again, making himself as flat as possible as he slunk across the top of the bunker. The tip of his tail swished back and forth in anticipation of the catch. Then he made a flying leap onto the back of a sibling and pretended to bite his spine, near his backside.

Elephant dung and sand flew as a chase ensued. In a burst of feline energy, the tackled sibling quickly got to his feet and ran after his attacker. When he got close enough, the pursuing cub reached a paw out and snagged the hind leg of his prized catch, causing him to tumble. The two cubs rolled around together, taking turns pinning the other down and giving each other huge play bites to the head and throat.

The play bout led to further antics, body contortions, swatting, tail-catching, tripping, and pileups. The other cubs joined in the fun, including mom, who instigated several more play bouts before suddenly running into the shallow pan. There, Paidia stood with all four legs submerged in the ice-cold water, as if inviting the cubs to join her for a dip.

As soon as the friskiest one approached, Paidia immediately charged with claws extended and teeth bared. She ran him out of the water with a smack on the rump as the others looked on from a safe distance, tails perked straight out with surprise and curiosity.

Again, Paidia stood and waited, and again, the same cub took the bait and received the same reprimand. This time, the charge was even more intense, leaving the risk-taker muddy and confused.

Although lions will cross a river if they need to—and I've seen many follow their prey into a shallow pan or even a river—they're not the swimmers that tigers are and often avoid water. It appeared that mom's play bout had turned into a lesson about the perils of water. Considering how close we were to the park border, Paidia could also have been imparting a lesson to her cubs about dangerous boundaries in general.

Shrinking natural habitats in Africa, and a scarcity of water sources in some places, make it hard on young male lions just beginning to explore and establish their own territories, all while trying to keep a safe distance from larger and dangerous adult males. We had already witnessed a number of deadly encounters where young males were chased away by older male lions that did not want to share access to water or to females. Inevitably, some young males break through the game fence and enter cattle farms, which often does not end well for the lions.

Whatever Paidia's motive, she clearly had a lesson in mind with her play that morning. As much as play might seem like a distraction from critical survival efforts, such as mothers providing food for their young, expending energy on play *is actually very important to physical and social development and even to survival.* This has been shown with horses: Researchers have found that, when mothers play with their young, both male and female foals develop better conditioning and have better survival rates.

Over the past fifteen years, I have watched the lioness Bobtail, Paidia's mother, encourage play as part of her family's development. Now Paidia was doing the same for her cubs. That morning, she was using play to hone their instincts of self-protection, even as the cubs were learning their own lessons while playing together.

What struck me most about the behavior of these cubs was how all of their postures and actions while playing replicated exactly the skills they would need as adults. In a lethal encounter with an adult male, they would use the same evasive tactics in self-defense. To catch and kill an antelope or other prey, they would repeat their early-morning antics of tripping a sibling with a paw, or running, pouncing, and biting the spine along the back or neck. Play-biting the esophagus was another favorite, which is what lions do to suffocate their prey.

GREETING RITUALS

Male black rhinos greet each other by crossing horns, akin to clacking longswords before a joust.

GREETING RITUALS

Top: Zebras greet by nipping and then grooming each other. Bottom: An adult male African elephant greets a higher-ranking male with a trunk-to-mouth greeting. This greeting can be a simple hello or a salute to a general. A particularly high-ranking male can inspire a line of greeters, akin to kissing the ring of a religious leader or Mafia don.

GROUP RITUALS

Top: Sailfish corral anchovies into a bait ball by encircling a school and holding their dorsal fin up, which forms a loose net. They also flash blue coloration across their bodies and fins, presumably to confuse their prey. The sword is used to draw a bait ball up to the surface to trap it, as well as to thrash through the middle of the school to weaken the fish, making it easier to catch them. Bottom: Female African elephants, Big Momma and her daughter Nandi, kneel down in a coordinated rescue of Nandi's calf.

GROUP RITUALS

Top: Young lions ambush a black rhino as they learn how to hunt. The hunting party needs at least one adult male strong enough to take the rhino down. Bottom: Lionesses take their hunting positions as they strategically plan and target their kill within a large herd of eland.

COURTSHIP RITUALS

Male and female giraffes wrap necks to court.

COURTSHIP RITUALS

The male ostrich sports a brilliant pink beak and shins due to high testosterone levels during the mating season. Courtship includes an elaborate feathery dance with his black and white plumage, bending his knees and undulating his long neck from side to side to impress his drab would-be mate.

GIFTING RITUALS

A gorilla mom cuddles her baby. The gift of touch and cuddling stimulates the production of oxytocin, the "bonding" hormone, making us feel warm and fuzzy. Pediatric wards hire cuddlers for babies whose parents can't be present as much as they'd like. Cuddling is a gift to both the giver and the receiver.

SPOKEN RITUALS

Top: African elephant family groups have to compete for access to water, particularly in the desert where water resources are at a premium. Sometimes altercations between families are quite vocal, including roaring and bellowing as the dominant family barrels aggressively in to displace a lower ranking family. Bottom: Male black rhinos are extremely territorial. If they don't time their arrival at the waterhole correctly, there could be an altercation resulting in a nightlong contest of bellows, heavy breathing, and bluster.

UNSPOKEN RITUALS

Top: The laugh expression of the chimpanzee, also known as the "play face," displayed while play wrestling and tickling, is similar to our own. Bottom: When in the hormonal state of musth, Smokey secretes a scented fluid from his temporal gland behind the eye, dribbles urine, and engages in a suite of exaggerated behavior to announce his status.

PLAY RITUALS

Paidia initiates play with her cubs, while also using play to impart lessons. Tripping, swatting, chasing, and biting the neck and spine are favorite lion cub activities that help them develop motor skills and agility for hunting as adults.

PLAY RITUALS

Elephants spend a lot of time playing. Very early they learn the ritual invitation to play: reaching their trunk over the head of a prospective playmate, similar to the play bow in dogs.

GRIEVING RITUALS

Lakota howls at the loss of his brother, Kamuts. Wolves howl for many reasons. We are most familiar with their howls at night, when they howl as a group to define their territory. They also howl for other purposes, such as to express grief. When the alpha male of the Sawtooth pack, Kamuts, died, Lakota howled every night for two weeks.

MIGRATION RITUALS

A herd of bison in Yellowstone National Park, starting to grow in their winter coat. Bison migrate from their higher elevation summer range to lower elevations in the

winter. Bison once roamed North America in large herds. They migrated through the Great Plains, moving south in the fall and north in the spring.

RENEWAL RITUALS

An adult male elk in Yellowstone National Park grows a full set of velvet antlers every year that will calcify in time for rut in the fall—when they compete with other males for a mate. The antlers are shed in early spring, when males have lower testosterone levels. They almost immediately start to grow new ones. Their high-pitched bugling is a spectacular sound of the wild, used to attract a mate and to assess the status of other males.

play rituals

This is one of the primary purposes of play. Play behaviors are, by nature, ritualized or exaggerated forms of routine behaviors that serve to perfect important adult skills, such as hunting, competing for a mate, or avoiding a predator. However, the play environment offers a special zone of protection. It allows one to experiment with any number of variables, including the element of surprise, without the potential consequences of the real world.

THERE ARE A number of different ways to categorize play, but most humans and nonhuman animals participate in *social, locomotive,* and *object* play. Oftentimes, play includes all three, such as pretend role-playing with others while incorporating objects and various forms of wrestling and chasing. Some theorists argue that games with established rules are not considered true play, since rules do not allow for improvisation or an outlet for free expression.

However, all play, in any form, abides by implicit rules of conduct and ritualized gestures, even if these rules are not explicitly prescribed on the back of the box of a board game. Take, for example, the invitation to play in elephants. A young male elephant might approach another while holding his trunk high above his head, presenting it toward the other in invitation. Juvenile male elephants engage in sparring behavior with male relatives of the same age, as well as with young males from other families. This serves to teach them how to fight, test their strength, and compete for a mate when they reach sexual maturity.

Adventurous male calves might stray far from their mother's side to initiate a spar with a distant relative. If the roughhousing goes too far, the calf will quickly run back to mom to seek her protection—sometimes trumpeting as he goes. In very young calves, the invitation to play is more of a pig pile or a mud-bath tackle—no formal invitation necessary. Sometimes these bouts are overseen by an older sister to make sure they don't get too carried away.

In dogs, the invitation to play usually takes the form of splaying the front feet and bowing low to the ground. This posture, known as the play bow, is an encoded invitation that says what commences from that point onward is pure play, with no harm intended. This establishes mutual trust, and if that trust is broken by a participant becoming aggressive with the intent to harm, the game ends.

In wolves, the play bow is often initiated by the omega. This is an important role, as play helps defuse tensions within the pack.

One essential element of play is experimenting with risk; the goal isn't necessarily to "win," but to practice and improve essential skills. Players actively seek unexpected situations by deliberately relaxing control over their movements, or they will actively place themselves at a disadvantage. Over the course of a play bout, the players will switch roles, sometimes taking control, sometimes self-handicapping. This allows everyone involved to test their strength and motility, to dominate, attack, and defend when compromised. Play researchers believe that the exploration of these modified actions leads to a versatility in motion and innovation, which, for example, would help one recover from a loss of balance.

Of course, locomotive play—running, walking, jumping, and pouncing—develops lifelong motor skills. Among all species, play fighting and roughhousing allow us to test our strength and agility among our peers in a safe environment. I see many examples of locomotive play on a daily basis among animals in Africa. Juvenile zebras break off into sparring pairs that nip and kick and call back and forth, practicing for when they will fight to protect their harem of females.

Giraffes play fight using a behavior called "necking," wrapping their necks around one another and then slamming their horn-like ossicones into the flank of the other. This activity simultaneously teaches them an important courtship ritual, where the slow and intimate wrapping of necks is a determinant of mating success.

Male rats also exhibit a form of play that is both practice for competitive fighting for a mate as well as social practice for courtship, including neck-biting. Even we humans sometimes incorporate playful and provocative sparring into our courtship flirtation.

When elephants playfully spar, it is akin to arm wrestling or even, when it becomes more elaborate and intense, like martial arts or jousting—both of which practice the skills needed for mortal combat. In addition to strengthening physical skills and reflexes, play fighting allows us to test boundaries between fun and threatening actions, as we learn what lines not to cross and how to read the body language of others.

Play is a low-cost, low-risk way to learn new behaviors, and the benefits are both immediate and long-term. Play is a particularly influential ritual for personal growth and confidence-building. Not everyone has the luxury of time or the resources to play, however. All forms of play, for both humans and nonhuman animals, peak during the juvenile stage of development, a period when there is typically more resources and protection. Then, in adulthood, the stress of survival and security, and the responsibilities of raising a family, can weigh on us, and play diminishes.

When animals are stressed or don't have enough to eat, or if the environment isn't safe, play is less likely to happen. This could ultimately contribute to a population's demise.

Play is far more than just a fun way to pass the time. It ignites innovation and exploration, and it promotes risk-taking and a flexible mindset for problem solving. Recent studies have shown that play is an important pursuit at any age, since it develops important skills and attributes we need to survive.

wild rituals

CONSIDER THE CAUTIONARY tale of *Homo erectus*. We don't know how much *Homo erectus* might have played, but maybe they needed to do more of it. Researchers suggest that they may have adopted a "least-effort strategy," and this led to a lack of risk-taking and innovation that may very well have led to their extinction.

During the early Stone Age, *Homo erectus* used tools that were adequate for the task at hand, such as for making other tools or collecting resources. They could have made better tools by making the effort to climb mountains to obtain better materials, but apparently this was too much work. Instead, they used whatever stones rolled down the mountains and were easy to reach. On the other hand, evidence shows that early *Homo sapiens* and *Neanderthals* climbed mountains to find quality stones for tools, and they transported those materials over long distances. This exploration and effort made them more equipped to deal with the coming challenges of a changing environment.

When the environment did change, and dried into a desert, *Homo erectus* did not adapt with it. There is no evidence they improved their tools, and researchers suggest that a combination of laziness and a lack of innovation likely led to their demise.

Play fosters risk-taking, which is a survival skill. A more modern example is a study that compared risk-taking behaviors between a traditional desert society, the Himba of Namibia, and Western societies. The study was simple. A person was asked to track a set of squares on a computer screen, following a simple set of rules. Clicking on the squares led to a slow accumulation that eventually achieved the goal. Every once in a while, a triangle would appear instead of a square. There were no instructions for this triangle, but clicking on it caused the goal to be achieved directly.

Because the Himba are accustomed to uncertainty—they live in an environment with minimal resources and a lot of fluctuations—they were more likely to click on the triangle, to try something out of the ordinary. The Western study subjects were not, perhaps because they enjoy a more comfortable environment, one that does not require innovation and continual adaptability.

Play has been shown to foster mental agility and overall brain health *at any age*, and not just in humans. In one study, researchers gave one group of mice, both young and old, a running wheel in their habitat. Another control group did not have a wheel. Within one month, older mice in the wheel habitat were faster at memorizing and finishing a maze challenge than the older control mice. This showed that locomotor play improved learning and flexible thinking. In addition, the brains of mice who engaged in play-like voluntary exercise showed an increased ability to adapt and transmit signals more effectively. On top of this, running resulted in better-quality growth of new neurons, promoting brain health. In other words: You *can* teach an old dog, or rat, new tricks!

Another study compared rat pups that were allowed to play to a group that wasn't. As adults, the play-deprived rats didn't react in the same flexible and fluid manner as those that were allowed to play. The lack of play impacted their ability to respond to new experiences and made them less successful when challenged.

IN HUMANS, PSYCHOLOGISTS have shown a direct relationship between playfulness and cognitive development, especially when we are young—and particularly with play's role in language development. When children play with peers, they create pretend scenarios. While playing "doctor" or "house," children are required to name objects, articulate what they are imaging, and negotiate with their friends. All this promotes language. Interestingly, their language is more varied and complex when talking to kids their age than it is when interacting with adults. This shows that playing in an unconstrained environment welcomes kids to explore and experiment even more so with their language and thoughts.

Parental involvement in play can also help develop a deep bond between children and their parents and other adults. Children learn parental boundaries, explore morals, and develop a sense of imagination by watching adults play. Play is also critical therapy for both children of addicts as well as recovering addicts, providing them with a safe outlet to rebuild confidence and strength.

The importance of play cannot be understated. Studies have shown that severe play deprivation in childhood leads to abnormal neurological development, and that play reduces symptoms of ADHD. Play therapy is a dedicated field, for both children and adults, that uses play in a protected space to explore creativity and one's own sense of self to foster mental wellness.

In her article "In Defense of Play," Alison Gopnik says that we don't play because we think that it will improve cognitive functions. We play because it's fun. Through play, we develop friendships and social connections, and strong social bonds lead to healthier, happier lives. However, Gopnik and others believe unstructured play is what's most beneficial. When children become overscheduled and lack "free time," this can undermine play's benefits, such as stimulating creativity and bonding. So can video games (and social media). Some researchers say that, when kids play video games together, this can strengthen bonds and be a form of positive play, but others warn that video games can be highly addictive and lead to isolation and reduced socialization.

Thinking about how play is, in essence, about having fun always reminds me of a four-year-old gorilla named Henry I once met at Zoo Atlanta. Henry's mother, Coochie, was considered a "helicopter mom" by her keepers. Coochie was overprotective and overattached to Henry, and she was depriving her infant of the opportunity to build relationships with his peers through play and other bonding activities.

It quickly became apparent how Coochie got her nickname. I was in an area that is off-limits to the public, and I was surprised to see an adult female gorilla allowing a very large male to suckle as if he were an infant. Henry was not yet an adult, but he was certainly too large to be nursing, yet his momma, Coochie, just couldn't let go.

In fact, Coochie kept an iron fist on Henry's leg and never let him go anywhere. This meant no play, no opportunity to explore, no interaction with others. This could spell disaster for Henry's healthy development.

I was fortunate enough to witness one of Henry's rare escapes from the shackles of his helicopter mom. As I watched, he cartwheeled around the enclosure, clearly hoping someone would join him in his play bout. Sadly, no one did. Nevertheless, Henry continued cartwheeling out of sheer joy and then started doing what his keepers called "air guitar playing" as he ran all the way across the enclosure. Finally, Henry could let loose and just have fun.

Embracing silliness is part of play. Being silly is not something frivolous that children should outgrow. Being silly is actually a highly adaptive behavior because it provides the opportunity to think far outside the box, to shake up routines, and to overturn convention just to see what happens. If we, as adults, embrace play throughout our lives, we will continue to nurture innovation, develop skills, and build relationships. Play is also a major source of stress release—from coloring books to squeeze balls, fidget toys, and Zen sandboxes. For all these reasons, many companies and corporations strive to create a sense of community and innovation through group games, role-playing, and corporate retreats.

MODERN ORGANIZED SPORTS are another form of play, and most have origins in the competitive games of traditional cultures. Ice hockey, lacrosse, football, baseball, and curling, to name a few, all originated out of the playful, often informal games of the past to become local, national, and even international institutions.

Despite their strict rules, organized sports evolved from and are considered a form of play and offer many similar benefits. They help individuals bond and build comradery; they give individuals an important opportunity to learn physical skills and express their prowess; and they provide a safe space to learn how to engage and defuse conflict.

In 1883, Baron Pierre de Coubertin of France, after visiting a rugby school in England, said he believed that the moral and social strength gained from organized sport was the reason for Britain's power. A decade after losing the Franco-Prussian War, de Coubertin believed that organized sport could help France create better soldiers, by fostering comradery and trust.

Inspired by these two motivations, the baron reinitiated the concept of the Olympic Games. They were originally held in Olympia, Greece, starting in the eighth century BCE but they ended in the fourth century CE. The first modern Olympics took place in Greece in 1896, bringing together previously warring nations in order to diffuse tensions and build alliances.

International sports competitions have bolstered national pride and reconnected fractured countries at very divisive times throughout history. In the 1980 Winter Olympics at Lake Placid, New York, the US men's hockey victory over the Soviet Union in a medal-round game served this purpose. I remember the "Miracle on Ice" clearly. My father had been a Russian major in college, and while in the US Air Force, he was stationed in a remote Maine outpost during the Cold War in the 1950s. At the time, the United States thought Russia would invade at any moment, and hence they valued having my father, as a translator, at the ready. So perhaps it was even more poignant for us children to watch our father's enthusiasm as the United States won this game and then went on to beat Finland and win the Olympic gold medal.

In America, the enormous sense of national pride that this game engendered was credited with helping to galvanize a nation still fractured after the Vietnam War. *Sports Illustrated* called the win "the top sports moment of the twentieth century." To the great shock of the Soviets, several Russian players later defected to the United States. After the breakup of the USSR, even more players joined the NHL. This historic game was later depicted in a made-for-TV movie called *Miracle on Ice*.

Another example of professional sports creating a healing moment for national pride was the notorious South African rugby game on June 14, 1994, when the Springboks played the New Zealand All Blacks in the nation's first post-apartheid game. President Nelson Mandela embraced the game, saying, "Sport is more powerful than governments in breaking down racial barriers." My husband and I watched the game with a group of friends in Namibia. A friend, who managed a wild game farm just south of Etosha, had us all over for a *braai* (barbecue) and rugby. The excitement for this historic game was palpable. We all knew that we were witnessing a moment that would be immortalized, owing to Nelson Mandela's vision for unity. When it ended, Nelson Mandela presented the winning cup to the Springboks, wearing a Springbok rugby shirt and cricket cap, and this gesture is credited with doing more to defuse tensions in postapartheid South Africa than any other single event.

HERE IS THE lesson for us: Play is never just about play. Both evolutionary and psychological evidence shows how important play is to humans. As a species, our highly adaptable, innovative, and social nature is rooted in and fostered by play. Play is not only good for our individual health but good for society, and providing time and opportunity to play should be considered an essential part of our agenda. Every day.

Of course, work is important, cooking dinner is important, along with all the other important tasks and obligations that seem to fill up all our time—but nothing is so important that we should sacrifice play. Nothing is more important than playing with a partner, a child, a pet, or friends, or even our partners, children, pets, and friends all at once. Play improves health and relationships and develops coping skills that help everyone, children and adults, meet life's challenges. During the coronavirus pandemic in 2020, my husband and I played more games (online) with friends than we ever had in the past, and this helped all of us reduce the enormous stress of that difficult time.

According to a long-term study at Harvard University, the best predictor of how healthy and happy a person is at eighty years old is not wealth or professional success, but the strength of that person's relationships at the age of fifty. Many of those middle-aged relationships began much earlier. Thus, friendships are lifelong pursuits that are initially shaped and can be strengthened through play.

Adults often consider themselves "too old" to play. They particularly don't engage in the unstructured play of childhood, the type of open-ended, improvised games that allow for real exploration, risk-taking, and innovation. Exercise and organized sports are great, but we should never forget the benefits of silly, on-the-spot, made-up games and the life-enhancing skills they teach.

I am grateful to Paidia—whom I named after the Greek goddess of play—for showing up that morning at Mushara to remind me that play rituals can serve as important life lessons. While watching Paidia initiate and fully partake in play with her cubs—and becoming just as committed to play as her cubs were—I realized that spending energy on play is just as important as any of

play rituals

the other many things we do for our health and happiness. As a society, our goal should be to incorporate as much play, and especially unstructured play, into our schedules as we can. Our survival depends on it!

RITUALIZED HEALING THROUGH TIME

> "I greet you from the other side of sorrow and despair, with a love so vast and shattered, it will reach you everywhere."
> —LEONARD COHEN

A SMALL FAMILY OF ZEBRAS had been standing in a cluster near the back of my field camp at Mushara waterhole for much of the day. Typically, the only wildlife that hung this close were the low-ranking resident rhino, which we affectionately nicknamed Scratchy, and some of my favorite male elephants that liked to watch our comings and goings, especially an elephant named Willie Nelson.

By the next morning, the zebra family hadn't budged. I thought it was a little strange, but hundreds of zebras visited the waterhole on a daily basis—so perhaps it wasn't so unusual for a small group to linger behind. It wasn't until one of the zebras within the family suddenly collapsed that I realized something was wrong.

When a zebra is lying down sleeping, it can look completely dead, legs sticking straight out as if rigor mortis has set in. We had come upon such slumbering zebras numerous times while driving up the unpaved sandy road to camp. They'd be lying in the middle of the road, stiff as a board, forcing us to drive off the path to slowly get around them. This would finally startle them awake, causing them to pop up on all fours and saunter off, swishing their tails and bobbing their heads—as if it were no big deal to return from the dead.

This time was different. As soon as the zebra collapsed, the entire family held their heads down and looked on as the zebra lay flat out and motionless. They seemed to know that this was no nap. It quickly became clear to me why they had stayed all this time—to stay with their sick family member.

Occasionally, one of the older females nuzzled the dead zebra and stamped a foot, another pawed the ground, and a few others bobbed their heads up and down. There was a whinny every now and again, but nothing—no response. The zebra had breathed its last breath.

Little is known about how members of the horse family respond to the death of a relative. In a group of garrano horses in northern Portugal, a study showed that a mother remained with a mortally wounded colt for an entire day, nuzzling him, whinnying and licking his wound, and trying to encourage him to rejoin the harem. Initially, the whole group took an interest in the colt, but in the end, only his mother remained at his side. Given the number of Iberian wolves in the area, it was not safe for a mother to be left defenseless. After persistent efforts by two bachelor males within the harem to retrieve her, she eventually returned to her family, where she would have had very strong bonds with the other mares.

Veterinarians who have had to put horses down have occasionally reported similar grief-like behaviors in horses. In a study of horse responses to euthanized companions, veterinarians described those companions acting possessively toward the dead individual, as well as withdrawing socially, not eating, and showing signs of anxiety.

Witnessing the zebra family's unwillingness to leave the body of their loved one made me think about how difficult it must be for all social mammals to make the hard decision to leave a fallen family member and move on. Thanatology, the study of death and the psychological and social conditions surrounding death, has traditionally focused only on humans. Now it includes some insects, birds, and many social mammals, particularly monkeys and great apes. A number of studies on social animals focus on the question of whether grief could be attributable to certain behaviors relating to carrying, attending to, and burying the dead—even the concept of mourning the dead.

Mourning behavior can be costly, both physically and psychologically. By remaining with the body of the deceased, grieving animals are more vulnerable to predators. In addition, their elevated stress levels take a toll on their overall fitness. Why, then, would animals evolve to grieve? B. J. King, in her book *How Animals Grieve*, suggests that the social withdrawal during grief could facilitate rest and recovery. This highlights why it is important for us to give ourselves enough time to reflect as well as to take opportunities to be alone to process our feelings.

Social animals respond to death in a number of different ways. The first suite of behaviors is called "mortuary"; these are mechanistic in nature and considered hardwired, or detached from emotion. Mortuary behaviors relate to practical issues of determining morbidity by checking vital signs; some species smell or lick the body, perhaps to identify the individual. The next step is to deal with the body, whether that involves carrying it to another location or burying it. Earliest documentation of human burials suggests the need to remove the body from proximity in case of potential disease and contagion.

Anthropologists make a distinction between practical mortuary behaviors and more escalated behaviors that involve some measure of emotional distress, or grief, called "funerary" behaviors. These tend to last longer. Researchers have put forward two criteria to determine whether a nonhuman animal's response to death is considered grief.

The first criteria is that two or more individuals choose to spend time together outside the survival behaviors of eating and mating. The second is that when one individual dies, the behavioral routine of one or more individuals around it changes. They might eat or sleep less or exhibit postures or gestures that indicate depression or agitation.

Many researchers believe that chimpanzees probably grieve in the human sense of the word (their brains show similar responses to emotions, including grief, as humans). They exhibit agitation and subdued behavior in the presence of dead individuals, both related and unrelated. This shows that they have an awareness of death within their own species. They also have similar responses to the death of family members as we do.

Not only do some species like chimpanzees exhibit the physiological markers of grief, but some exhibit extreme grief, based on long-term studies of pathological reactions to loss. In Jane Goodall's chimpanzee study in Gombe, she observed a chimpanzee named Flint from birth until he was eight years old. Flint was overly dependent on his mother, Flo, the matriarch of the group. When she died of old age, Flint withdrew from his family and remained near his mother's body continuously until he himself died within a month.

Studies show that wild female chacma baboons display signs of grief at the loss of a family member. During one study, a baboon family member was killed by a predator, and the rest of the females in the family had very high stress levels. Stress was measured by the amount of cortisol, a stress hormone, found in the stool of each individual. Unrelated females within the troop did not have higher stress levels. After the death of the family member, the bereft individuals also spent more time grooming with a greater number of partners than normal. The researchers suggest that, by extending their social network, the related females might have minimized the stress they were experiencing from the loss. Two months later, their stress levels all went back to normal.

Cortisol levels also go up in humans during bereavement, and by extending social networks during mourning, cortisol levels and stress go down. Just as we do, other social mammals like baboons also rely on an extended community to soothe the loss of a family member.

Social insects—such as ants, bees, and termites—also exhibit mortuary behaviors. Once a number of individuals inspect the body—having been alerted to the death by a hormone, called "necromone," that is given off by the dead individual—a specialized undertaker in the colony removes the dead body or buries it within the colony by walling it off. Necromones are fatty acids in insects associated with decomposition and are a strong signal for members of the same species to stay away—possibly to protect the others from something contagious. It's not only insects that secrete the smell of death. Sharks can also detect necromones in other sharks. Even humans have putrefaction volatiles that act as necromones. Some researchers suggest that these necromones are the signals that trigger a dog's ability to identify diseases in humans, a species that they evolved in partnership with for thousands of years. This ability could help both species survive.

In one giraffe family in Kenya, researchers observed that a mother gave birth to a baby with a deformed foot. The calf was not able to keep up with the rest of the family and only lived for a month. During that entire time, the mother remained within sixty feet of her calf until it died. An hour after its death, seventeen females from the family inspected the body and remained within the vicinity. Several hours later, twenty-seven giraffes had amassed to attend and inspect the body. Even through the next day, a number of adults remained with the body, including the mother. By the fourth day, the body had finally been taken by hyenas.

While the mother remained the entire time, the prolonged attendance of additional family members suggests they were protecting the dead calf from predators. Another interpretation is that the mother-calf bond was so strong that it was very hard for the mother to leave. Since she couldn't carry her dead calf, she remained with it, as did other family that shared a close bond with the mother. Some researchers believe that the mother-offspring bond after the death of a calf is even stronger if the calf lived long enough to suckle. The so-called "bonding" hormone, oxytocin, plays an important role in lactation and would have been secreted in the mother, creating an even stronger bond and making it even harder to leave her now-dead calf.

Grief in wolves is manifest in different ways, but the most compelling observation is that no matter which wolf has fallen from the pack, the rest of the pack experiences the loss. During the early days of the Sawtooth pack, the first low-ranking female was killed by a mountain lion. When she died, the rest of the pack stopped playing and did not howl as a pack for a period of six weeks. On the occasion that one or another howled on their own, they did not stand in the typical howling posture, and the howls took on a different tone. To the researchers that were accustomed to hearing their nightly chorus, they sounded mournful.

The death of this low-ranking female wolf illustrated that, despite being ostracized and chased by the others, the omega is still an important member of the pack. When the alpha male, Kamots, died, his brother, Lakota, howled every night for two weeks.

In their book *The Wisdom of Wolves*, Jim and Jamie Dutcher write about another wolf pack in Alaska's Denali National Park. A researcher reported that the pack's alpha female had been caught in a hunter's trap just outside the park. During the two weeks she was trapped, the researcher believed that the alpha male and their litter from the previous year brought her food. When the trapper finally returned and killed her, the alpha male traveled fourteen miles back to the den where their previous pups had been born, cleaning it out for a litter he would never have with her. He then traveled back to the area in which she had been killed. The researcher saw him howling in the direction of the trap line and could only attribute the alpha male's behavior to bereavement.

Learning about this wolf not wanting to leave the location where he last encountered his partner reminded me of my experience watching elephants searching for a lost companion. Late in the field season of 2013, the bulls Johannes and Keith arrived at our camp just as the waning moon was rising yellow above the horizon. There were only three of us left in camp that night, as most of the team had recently left.

I wanted to take advantage of the quiet to do playback calls of "let's go" rumbles from known elephants to see if their bonded companions would recognized these calls and respond. When Johannes entered the clearing with Keith, I wondered how they might respond to a call from their very close companion and the dominant bull at Mushara, Greg.

I waited until Johannes and Keith had a drink and were on their way out, walking up the northeast path to the left of the tower and back into the forest. As soon as they reached the edge of the clearing, I broadcast Greg's signature "let's go" rumble, and the two immediately froze. They stood there for some time and I played the call a second time.

I will never forget what happened next. The two of them turned around and came back to the waterhole, rumbling in response to hearing Greg's call. Then they started searching. They placed their trunks on the ground, each of them facing a different compass direction and standing close to each other in silence. A few minutes later, they rumbled again, repositioned, and placed their trunks on the ground in another direction, and stood in silence again.

After repositioning four or five times, I started to feel terrible. We hadn't seen Greg for more than a year. What if his friends hadn't seen him for that long, and were genuinely searching for their long-lost buddy?

The experience haunted the three of us as we watched this heart-wrenching response. If Greg had indeed passed away, I wondered if his buddies had been nearby when it happened. If Greg had died, how long would his companions have attended him? These are the questions that I asked myself as I lay in my tent looking up at the Milky Way, listening to the rise and fall of Johnannes and Keith's deep low rumbles through the night in search of what may have been Greg's ghost, shedding tears for the fallen don of Mushara. I couldn't replay Greg's call again.

There are many reports of elephants refusing to leave a dead family member. In one case, a mother awkwardly carried the stiff body of her calf around by the trunk for some period of time before making the decision to leave the baby behind. Carrying behavior has been documented in a number of species, including apes, monkeys, dolphins, and dingoes. In the case of dolphins, mothers have been seen carrying their dead baby along using their dorsal fin. In one instance, a dolphin mother carried her baby for five days while being escorted by close associates and an extended family. On day five, two of her closest associates also assisted in carrying it.

Carrying usually lasts from one to seven days with the mother exhibiting grooming and protective behaviors as if the baby were still alive. In one small chimp population in Guinea, an experienced mother carried the mummified remains of her baby for almost seventy days. In nonhuman primates, we might think mothers might carry their dead out of confusion, but researchers believe that these mothers are aware of death because they carry the body differently than if their baby were alive.

A number of explanations have been put forward as to why carrying the dead might persist. As with the giraffe mother, one is hormonal—once the mother stops lactating after the baby dies and enters postpartum, her protective instincts kick in and she has a strong urge to hug her baby. Another potential explanation is that the carrying behavior signals to attractive males that the mother is protective and would be a good mate.

The instinct to carry, hold, or remain with the dead may have important physical and psychological benefits. A recent human meta-study combining twelve studies from six countries, including the United Kingdom, the United States, Canada, Australia, Sweden, and Japan, reported parental views regarding their experiences of stillborn births. The study spanned four decades, revealing many different hospital practices. These ranged from hospitals not allowing bereaved parents to see their stillborn child to parent organizations encouraging healthcare professionals to allow a visitation if the parents requested it.

Many parents expressed the importance of seeing their stillborn child as a way of coping and mitigating depression by generating a memory with which to grieve—such as holding their wrapped baby and seeing his or her face, even for a few moments. When parents declined to see the body, some were given a box of relics containing mementos of the child, such as an ink imprint of a hand or foot, or a hospital band. Whether the parents chose to open the box or not, many reported that knowing that they had these items in their possession was a great consolation.

Given such powerful experiences of bereaved parents, it's reasonable to expect that chimpanzees would have similar instincts. Whether it be conscious grief or purely for hormonal reasons, the two are likely inextricably linked. Holding or carrying the dead allows the bereaved to spend more time with their loved one, providing more opportunity to grieve, and thus perhaps a healthier outcome for the bereaved. This would suggest that grieving is, in fact, a survival instinct.

MANY ANIMALS CAN'T carry their dead, except sometimes with small infants. This is likely why, as was the case with the bereaved zebra family, zebras will remain with a deceased family or group member for long periods of time before leaving the area, most likely in order to eat and drink before returning to the body. Meanwhile, adult elephants have very large bodies that remain in the environment for months to years, and individual elephants have occasionally been observed returning often to their dead relatives or to the site where their relative died.

I watched elephants returning to the body of a dead relative during an anthrax outbreak among the animals in Etosha National Park. When an elephant is sick or wounded, it doesn't stray far from water, so elephants often die next to a river or waterhole. At Rietfontein, one of my favorite waterholes along the drive up to our field site within Etosha, elephants went out of their way to walk next to an elephant that had recently died. It wasn't just family that stopped to smell and touch the body on their way to drink, but several different families visited the carcass—perhaps extended family, perhaps curious bystanders.

Researchers suggest that the length of time elephants spend visiting the remains of another elephant indicates more than mere curiosity, such as finding out who passed away. Females secrete fluid from their temporal glands either under duress or during a reunion, and temporal-gland streaming is observed during these visits. This altered physiological state suggests they are feeling strong emotion, which perhaps means these visits are on par with funerary rituals.

Elephants have a strong instinct to protect those who are struggling. I have witnessed family members close in around a faltering individual; they use their bodies to hold the individual up and keep them on their feet. Sometimes they even use their tusks in an attempt to prop them up. After death, elephants sometimes try to pick up the individual by using their front or back feet to lift them off the ground.

Vernon Presley, curator of elephants and ungulates at the Fresno Chaffee Zoo, has witnessed seven elephant deaths in captivity in his twenty-five years of working with them. In his experience in a captive setting, Presley says elephants exhibit different responses depending on their relationship to the deceased as well as on their previous life experience.

grieving rituals

On one occasion, a matriarch in captivity had to be put down. The matriarch had a persistent foot wound and couldn't move around normally. Eventually, her increasing immobility led the zoo staff to decide that ending her life was the most compassionate choice. However, she wasn't sick, and Vernon believed that her two closest companions would not be expecting her to die from her ailments. As such, the zoo staff were not sure how these two would respond to her sudden death, and they felt these two elephants would need an opportunity to mourn, if they wanted to, rather than wake up one morning to find her gone, with no opportunity to say goodbye.

The staff decided to leave the matriarch's body out in the exhibit, where the other elephants could visit her body. They didn't know whether the elephants might react aggressively to the dead body, but it was a risk they knew they had to take.

When the time came, several of the low-ranking individuals did not go near the dead matriarch. Two others sniffed at her and left the vicinity. But her two closest companions reacted much differently. They stood next to the body and explored her together, touching and smelling her. They took turns quietly visiting with her body throughout the night, never leaving her alone, each one sprinkling dirt on her periodically.

By the next morning, her whole body was covered in at least a quarter inch of dirt. This was the most compelling burial ritual Vernon had ever experienced.

One particularly striking detail in Vernon's story was that these two elephants were born in Mozambique and were captured and brought to North America at six and seven years of age. It is quite possible that these elephants had experienced the death of other family members in the wild and had participated in mourning rituals. When the matriarch died, the two other females, who were very closely bonded with her, may have remembered back to their youth and re-created the burial rituals their families performed.

Vernon thought this was a compelling possibility. He has not witnessed any other elephants in captivity replicate this ritual for a dead elephant, yet none of those elephants have come from the wild. They would not have had the opportunity to learn this behavior, which may have been passed down between generations.

Numerous accounts describe wild elephants burying dead elephants by sprinkling dirt on the body or covering it with branches. I have heard of cases where an elephant will cover a wounded person with branches, whether to protect them from lions or as a form of burial.

Wild chimpanzees bury dead family members, either in shallow graves or by covering them with leaves. In one case, researchers witnessed an individual fall out of a tree and die, and afterward, a group sat in the tree above the dead individual, broke off branches, and dropped them to cover the body.

Many social animals have pragmatic mortuary rituals that involve the removal or covering of the dead for practical purposes, such as avoiding disease, parasites, or some other contagion. These mortuary rituals are presumed to be the origin of human burial rituals. In addition, many social animals, like elephants, chimpanzees, and zebras, display funerary rituals, in which they seem to mourn for the dead and express grief in ways that are strikingly similar to us.

However, early humans took burial further still. Starting around four hundred thousand years ago, Paleolithic archaeologists find the first evidence of funerary caching, or burying objects with the dead, and about three hundred thousand years ago, they find evidence that people began setting aside particular landscapes for the dead. Both of these actions are considered the first evidence of an understanding of an afterlife.

This concept emerged as human culture and cooperation blossomed during the Pleistocene era, and over time, humans developed ever-more-elaborate treatment of the dead. Loved ones were buried

grieving rituals

with treasures and objects to accompany or protect them when they crossed over, and burial sites were often meant to be enduring, permanent monuments, like the Egyptian pyramids. Today, many people commemorate the dead in ways that reflect a belief in the existence of a soul.

OF COURSE, WHATEVER one's beliefs, grieving is hard, and we often struggle with how to express our grief. Often, we don't realize how important the process of grief is to our well-being, and we do what we can to avoid it. Grief serves as a time to reflect. Being present as a loved one passes is an important aspect of saying goodbye; it's an important way we accept their passing. Some form of group grieving just after a death is also an important stepping-stone to acceptance. It's no wonder that there are so many rituals around death and dying.

Watching that poor zebra family reminded me of my own family when they watched my brother pass away at his bedside. I was not able to be there with them. On my way to say goodbye, one of my flights was delayed, and I ended up missing the moment of his passing. I also missed being together with my family in the hours leading up to his death.

Just as the plane landed, I received the news; my sister sent me a text that my brother was gone. By the time my sister arrived to pick me up, my brother had already been removed from his hospital room, and everyone else had gone home.

Not being there to say goodbye in person, I found myself at such a loss. I was at unexpected loose ends with my grief, and it made it harder for me to process my brother's passing. His death seemed surreal to me. It made me realize how death is more alienating if you are not present to witness it.

Later, I was very grateful that the casket was opened to the family prior to the wake, when the casket would be closed. I needed to see my brother again, even under such stark circumstances. The moment I saw him lying there—peacefully, yet decidedly cold and still—I must have startled everyone with my sudden outburst of emotion.

I now appreciate why some cultures throughout history have engaged in the practice of hiring professional mourners, called "moirologists," to help families through the grieving process. In Egypt, China, the Near East, and some Mediterranean cultures, professionals can be hired to attend wakes and engage in grieving, lamenting, even eulogizing, as well as providing comfort to the grieving family. More recently in the United States, what are called "death doulas" serve as facilitators and grief counselors to the dying and their families, in addition to being bereavement specialists.

Researchers suggest that mourning rituals serve to support as well as contain strong emotions. Mourners come together with the grieving family for a short time in order to facilitate the initial expression of pain and loss in a safe community space. This atmosphere encourages communication and openness during the grieving process. Holding a wake, a funeral, and a burial, and then sharing a meal afterward, gives the bereaved time to process death and their new reality over the course of several days.

This window of time also gives the community the flexibility to pay their respects in a way that is most comfortable for the family. Shared meals and visiting serve to prevent the bereaved from experiencing a period of isolation surrounding their loss.

If people don't engage in mourning rituals and carry on with life as normal, they miss the opportunity to heal through grief. Meanwhile, other cycle-of-life rituals such as weddings, births, and holidays can serve as moments to remember the loss and help process grief. These events help to secure the memory of the deceased in everyone's hearts as their lives move forward.

In retrospect, my experience made me realize how important it is to grieve as much as possible *in the moment*, and if the moment is lost or we put it off, it may end up doing much greater harm. Studies have shown that people who suffer from persistent symptoms of grief are at higher risk of getting cancer, heart disease, and hypertension. It's best to face grief in the moment and allow our community to help support us. Grieving with an extended community helps to reduce stress and allows us to mature with grief constructively over time.

Another difficulty or obstacle to embracing grief is when others are uncomfortable with it. Death can be uncomfortable for many, yet death is a constant. It can take a survivor a lifetime to process the death of a loved one, and this happens in stages—sometimes being alone is important, and sometimes grieving as a group is especially helpful.

Group grieving is important, not only psychologically, but physiologically, as it has been found to lower stress levels. This includes societal grief. The magnitude of this has been brought home to me on two occasions: during 9/11, and in 2020, during the COVID-19 pandemic, when the *New York Times* listed the first hundred thousand Americans to die from it. But there are many examples of public commemoratives that foster the healing power of group mourning on a societal and even global scale: the Peace Memorial Park in Hiroshima; the Kigali Genocide Memorial in Rwanda; the Memorial to the Murdered Jews of Europe in Berlin; the site of the Twin Towers in Manhattan; Sandy Hook Elementary School in Newtown, Connecticut, and more.

Our capacity to feel grief for the loss of strangers evolved as a survival skill. There is something very cathartic about the concept of expressing grief as an anonymous collective.

Our fear of grief is similar to our fear of death. Yet anthropologists believe that prolonged interactions with the dead accelerate the grieving process. They also serve to restructure a community based on the altered relationship to the deceased. This has led to more elaborate rituals centered around the dead in some cultures, where grief and mourning also include a celebration of life.

wild rituals

In Mexico, the Day of the Dead is a multiday celebration of the lives of departed loved ones. Families bring food to their relatives' graves, and they believe that their ancestors join them in celebrating life. In China, Tomb Sweeping Day celebrates the dead by cleaning out their tombs. On All Souls' Day, the French visit their deceased loved ones as a family, bringing flowers or doing a reading in their honor.

These cultures teach us that expressions of grief are best balanced with joyful memories and gratitude for the person's life. By embracing the dead, we grow in our new relationship with them. It's a healthy practice to spend time honoring the memory of our loved ones.

When I arrived at my brother's burial site, I experienced a strange sense of peace, like I was visiting his next home. As he was lowered into the ground, however, I realized I still wasn't ready to leave him. The finality of the moment gripped me.

Sprinkling dirt on his casket gave me little solace. Then I thought about the elephants covering their matriarch with dirt, and the chimpanzees covering a family member with branches. These rituals connect us to all other social animals and their grief. As my husband and I slowly walked away through the pastoral setting, dotted with attractive headstones, adorned with fresh flowers, I tried to be happy for my brother's place of rest.

After writing this chapter, I went for a run on the beach to clear my head. It had been a hot day with a clear blue sky. The evening finally brought a chill to the air. It was a completely different scene than my previous run, when a cloud bank and gnarled surf sandwiched me against the cliffside at high tide, the day before full moon. Everything had been a luminous, but ominous gray. I could barely see beyond fifty feet.

The beach can have such a different character, depending on the weather and tide. Sometimes, it reflected my mood; other times, it shaped it. The harsher the conditions, the more inwardly focused my thoughts tended to be.

grieving rituals

The calm conditions on this night caused my thoughts to scatter—I was trying to get my footing and searched for an inner peace, but there were too many distractions. Finally, I caught hold of something.

Just as I reached my runner's high, I noticed many people had quietly gathered to watch the sunset over the Pacific Ocean. It was one of those perfect ones—the horizon was devoid of any cloud bank in the deep orange sky. The giant yellow sun slipped into the water, the last little crescent flashing green before disappearing. The sandstone boulders where everyone was sitting were lit up in a pink glow in the stillness—not unlike tombstones in a cemetery.

The impromptu crowd paying their respects to the sun reminded me of a group of mourners looking for a similar sense of peace. It wasn't somber. Rather, it was a celebration of life and the passing of a day marked by the setting sun.

As I left the beach, I was overcome with a wave of grief. This triggered me to sing the Cyndi Lauper song "Time After Time." I often turn to music when I feel overwhelmed, though I learned only recently that this has been shown to be very effective therapy for grief. Thinking of my deceased brother, the next song on my mourning playlist was Stevie Nicks's "Landslide."

Tears streamed down my face. Some were tears of loss, but some were tears of joy for all the memories. It had been almost four years since he had passed.

Humans are a remarkably resilient animal. It is through processing grief and practicing grieving rituals that we are able to heal.

… 9 · RITUALS OF RENEWAL

SPRING CLEANING AND OTHER RHYTHMS IN NATURE

> "There is something infinitely healing in the repeated refrains of nature—the assurance that dawn comes after night, and spring after winter."
> —RACHEL CARSON

THE TRANQUIL WATER SUDDENLY EXPLODED with giant black projectiles all around us. Breaching whales erupted from the depths, launching twenty feet into the air, dwarfing our sailing vessel and even the island of Lanai in the distance. When they were three-quarters out of the water, the whales rolled sideways and splashed back down into the ocean, disappearing from sight and into the deep blue depths below—only to relaunch a hundred yards ahead a few moments later.

In the distance, all the tourists on the whale-watching boats were oohing and aahing with every breach. It was a few days before New Year's Eve, and this was clearly nature's version of fireworks. Tim and I watched in awe from the deck of his brother's catamaran, floating between Maui and Lanai.

The humpbacks had returned to Hawaii from Alaska on their annual migration. It was calving season, and they had much to celebrate. As a pod of whales approached our boat, my husband, my brother, and I grabbed our masks and snorkels and jumped into the turquoise water. Even though I willingly jumped off the boat into the infinite sea, I couldn't help but feel a wave of anxiety to be in the water so close to the Earth's largest creature for the first time.

Underwater, I took a quick look around, but I saw no whales—just blue. I breathed easier knowing that I could just enjoy the refreshing water with my head submerged, listening to the sounds of whales calling all around us, without the fear of being in proximity to a sixty-six-thousand-pound living being.

Deep in the water, off to the right, a school of small fish caught my eye. Suddenly, a gigantic dark form took shape behind the school of fish. It was hard to judge the scale of what was heading straight for us from below, but it looked like a submarine—only it had long, white, flexible fins.

As the fins slowly beat toward us like the graceful flapping of wings, I could see that it was a humpback whale. My first-ever underwater sighting of one of these spectacular creatures. As the whale got closer, I saw a much smaller set of fins underneath her—a calf!

I dove down to get a closer look as Tim followed them into the depths with his underwater video camera. I took a picture of him dwarfed by that enormous, iconic tail as it disappeared into the dark blue distance. The majestic presence of this mother and calf was humbling. It reminded me of the importance of having these magnificent animals on this planet. It also reminded me of the importance of seasonal rituals of renewal, and sparked my own journey of renewal as it was the beginning of my fiftieth year.

Traveling three thousand miles from Alaskan waters, the majority of humpback whales arrive in the tropical waters of Hawaii between the months of January and March to calve and mate. The return of these whales is a unique indicator of the passage of time in Hawaii. Since this remote paradise lies near the equator, very little changes as the calendar turns. The average temperature ranges from seventy-eight degrees Fahrenheit in the winter to eighty-five in the summer. The sun sets at almost the same time all year round.

One seasonal shift is the onset of the Kona winds, which blow during whale season in winter; to locals, this means that surf's up on the north shore of all the islands. The milder trade winds during the summer months mean that novice surfers can flock to the south shore of Waikiki to score a gentle ride on a long board. The minimal swell also means a flat north shore with great visibility, so scuba divers head to Shark's Cove on Oahu.

Animals have a strong sense of the seasons and understand the passage of time, whether it be measured in the seasonal availability of food, the need to migrate or hibernate, the urge to court and reproduce, or a birthing season. Animals that live in seasonal environments need to adjust their behaviors to match the season. Their survival depends on it.

In temperate climates, those north and south of the equator, birds use the length of day as an environmental cue to track seasonal changes, which trigger important behaviors. Birds have extra photoreceptors in their eyes that are tuned in to light cues; these let the bird know when days are getting shorter or longer. For instance, as days get longer during spring, birds know it's time to mate and lay their eggs. If a bird lays their eggs too late, and they hatch at the end of summer, there may not be enough food left to eat and their young might not survive the winter. Further, if birds are moved as part of an experiment, their hormones and behavior will change to match their new environment. For instance, if Japanese quail are translocated from a place with shorter days to a location with longer daylight hours, their hormones will adjust within hours of arrival. In this case, since longer days signal that breeding season is approaching, the quail will be stimulated to develop larger gonads to prepare for reproduction.

The rufous-winged sparrow of the Sonoran Desert needs a different internal clock than birds living in a temperate climate. Since there is no food in the desert until after the monsoon rains of late summer, this sparrow doesn't nest until then. They need to change their sensitivity to daylength in order to lay their eggs at just the right time, but they still rely on light triggers, just like the Japanese quail. Changes in the environment stimulate hormone expression, which triggers a bird's behaviors, making the ability to tell time and detect a change of seasons a matter of life and death.

Mammals have the same hormonal mechanism for detecting a change of season. A bear needs to eat enough during the fall in order to survive a hibernation-like state called "torpor" during the cold, winter months. Since fall is when salmon return to spawn in the streams where they were born, this makes them an important part of a grizzly bear's diet. They arrive when bears most need to eat.

In fact, bears go through a physiological change called "hyperphagia" at this time, causing them to eat nonstop, even up to twenty-four hours straight, before sleeping for four and then gorging themselves again, sometimes only eating the fat-rich eggs, skin and brain of the salmon. In this way, they can take in on the order of a hundred thousand calories a day and gain the hundreds of extra pounds needed to survive their winter slumber.

In the Pacific Northwest, salmon themselves follow environmental and hormonal cues that regulate their whole lives. Salmon hatchlings emerge in winter and feed off their yolk until they become "fry" the following spring (and eat plankton). By the end of summer, the juvenile salmon are called "parr"; they remain in the rivers for several years until they become smelt—the stage in which they become salt-tolerant and head to the ocean, where they spend their adulthood. Then in late summer to early fall, magnetic cues in the ocean prompt adults to return and spawn in the more-protected environment of their birth rivers—and the cycle starts over.

After five molts, a monarch caterpillar metamorphoses into a butterfly and flies south and west in the fall to avoid the cold winter. Seals molt in the spring to shed their winter coat for summer. While they molt, they are vulnerable to the cold water, so staying out of the water during that time is important. Elk and moose males grow enormous antlers between April and August in preparation for the fall mating season, when males must compete for the females.

The first rain in an environment can be an important signal for many animals. Just after the first rains, termites have a nuptial flight where they burst out of the nest in a volcanic flurry of long brown wings with the mission of mating and starting a new colony. Every creature that eats termites awaits this moment. The migratory, southern carmine bee-eaters fly over the Zambezi River and enjoy a termite meal on their way to equatorial Africa to overwinter from March to August. Anteaters, honey badgers, and pangolins also join in the feast.

LIKE WHALES, BIRDS, and other animals, human circadian rhythms (our daily physiological functioning) are affected by the same day-length and light-exposure factors. Just being in nature, studies have shown, reduces stress levels, lowers blood pressure, decreases muscle tension, and lessens negative emotions like anxiety and depression. Experiencing and being aware of the cycles of nature—feeling connected to our environment in all the ways that plants and animals are—is beneficial and healing. However, whether it is because we are absorbed in our technology-driven life or too busy to spend time outdoors, many of us have become disconnected from the rhythms of nature, and we neglect to slow down to enjoy, to celebrate, and to rest.

From planting in the spring to harvesting in the fall, humans have many activities and rituals that respond to the signals of nature. *One of the most important seasonal rituals is that of renewal and new beginnings in the spring.* The mood-boosting aspect of spring is caused by the onset of warmer weather; an increased exposure to light also triggers dopamine production, which not only makes us feel good, but also improves confidence and memory. Growth hormones even kick in during spring.

Many rituals surrounding the onset of spring begin on the spring equinox, or the day when the sun is exactly above the equator. In the northern hemisphere, this alignment happens every year on March 21, when the sun moves north across the equator. The fall equinox, the day when the sun crosses the celestial equator moving south, occurs every year on September 22. The seasons are reversed for the southern hemisphere.

Rituals surrounding the spring or vernal equinox have been around for centuries. Many celebrations have their roots in aspects of Ostara, an ancient Celtic and Saxon holiday designed to worship Eostra, the goddess of dawn and fertility. For Wiccans and other Pagans, Ostara celebrates new growth and the marriage of Mother Earth with the Sun God. Christian Easter and Jewish Passover are observed at this time as well, the word Easter having been inspired by Eostra. Ancient Persians celebrated with a thirteen-day festival following the first day of spring, now called Nowruz, which means "the new day" and represents the Persian New Year. In Roman mythology, the god Mithras was resurrected on the spring equinox and created the moon and the night sky by sacrificing a white bull and his cloak. Mayans performed rituals at the temple of Kukulcan, a step-pyramid in the center of Chichen Itza in the Yucatán structurally designed to align with the long shadow of the afternoon sun for the equinox. Given how many cultures recognize the spring equinox, it's clear that how we once mirrored environmental cues in nature and celebrated their importance.

Ostara rituals, both ancient and currently practiced, typically include setting up altars decorated with spring colors and statues of gods and symbolic animals, along with baskets of eggs all symbolizing spiritual renewal. A man and a woman act out the courtship of the spring god and goddess and symbolically plant seeds, while activities involving eggs—races, hunts, eating, and painting—are engaged in as a group. Hot cross buns are a favorite symbolic food at Ostara, representing many different themes, such as the elements (earth, air, fire, and water), or the four directions, the four phases of the moon, or even the four seasons.

Just as the sun has a day where it is closest to the equator, the opposite is also true. The summer and winter solstices occur when the sun is at its greatest distance from the celestial equator, marking the longest and shortest days of the year. In Wiltshire, England, Stonehenge is aligned like the Mayan structures, but instead of marking the equinox, it marks the sunrise of the summer solstice and the sunset of the winter solstice. *Solstice* is a Latin-based word for "standing still," which the sun appears to do at noon on both solstices, since it is at the greatest distance from the equator.

Just as the arrival of spring and increased sunlight influence our mood, so does the arrival of winter, with its shorter days and less daylight. This reduced light exposure is the cause of seasonal affective disorder, when the lack of light can cause depression.

Researchers have suggested activities and rituals that help us acknowledge seasonal changes are helpful ways to reconnect us to nature and the natural cycles of life, giving us a sense of control over our lives and restoring a sense of well-being. Spending time outdoors can help connect us to the seasons. Gardening is one such activity. Bird-watching is another. Allowing oneself to experience these cyclic changes offers opportunities for renewal.

Documenting the timing of events within a journal, such as the first arrival of migratory birds in the springtime or the first hint of the leaves changing color, in autumn, can be therapeutic and has a long history. A record of the first cherry blossoms in Kyoto goes back to the ninth century. This kind of record-keeping can be done in one's own backyard, in a local park, or on a nature trail.

SPRING CLEANING IS another important ritual of renewal, and you might be surprised to learn that animals engage in spring cleaning like we do. Dusky-footed wood rats, more commonly known as "packrats," are known to place bay leaves in their nest as a fumigant to reduce parasites. These cute cinnamon-colored, bushy-tailed creatures with large round ears are very resourceful at cleaning house.

The nocturnal beach mouse does a bit of spring cleaning of its own. As the weather warms in spring, they discard the old husks of seeds and the exoskeletons of insects that they consumed over the winter, placing them outside the entrance of their burrow. The house mouse engages in the same behavior during March and April, when the entrances to the burrows become littered with removed debris and grass. Researchers believe that burrow-cleaning serves to remove older decayed material from the winter that may contain harmful insects.

Birds that reuse nest sites, like starlings, also "clean house," often by incorporating fresh green leaves containing volatiles into their nests. These chemical compounds exterminate or slow the growth of parasites, such as by disrupting their reproduction. Parasites can survive long periods in old nests, even lasting through a winter freeze, and the special leaves protect the birds' chicks from losing blood to infesting insects.

The honeybee hive is apparently one of the cleanest and most sterile environments in nature. Worker bees work hard to prevent disease by sterilizing the hive with an antimicrobial agent called "propolis" that they collect from trees. This substance is used to kill parasites or isolate a large intruding creature that they have killed but can't remove. It also provides a kind of social immunity to all the honeybees, which lessens the need for each individual bee to have a strong immune system.

Cleaning rituals are especially important in large animal populations where waste removal minimizes the possibility of contamination. There are many cases of nest-cleaning in social insects, particularly in termite and honeybee colonies that have designated underground areas that serve as refuse dumps. The leaf-cutter ant uses either a designated underground location or places its refuse in a pile right outside of the entrance to the subterranean colony. The pile consists of degraded leaf material, fungal biomass, deceased ants, and other colony waste.

To clean themselves of pesky parasites, many animals seek help from others. Cleaning stations are well known among fish communities, such as cleaner wrasse that occupy a specific section of reef and advertise their services with a ritual posture. These small colorful fish will suspend themselves upside down, waiting for their next customer to hover in place while being cleaned. Popular cleaning stations can have a line of fish waiting to be cleaned.

Some researchers believe that the origin of spring cleaning in humans stems from the Persian New Year (Nowruz), where everyone does a thorough housecleaning or what is referred to as "shaking of the house." From drapes to furniture, a deep clean to remove parasites occurs just before the new year starts, which for them is the first day of spring. If rituals to remove parasites from your home seem strange today, think about our modern bedbug problem. Keeping a clean house isn't just to remove dust!

Others believe the spring-cleaning ritual is attributed to the ancient Jewish practice of thoroughly cleansing the home in anticipation of the weeklong celebration of Passover, which is followed by a traditional hunt for chametz (leavened) crumbs by candlelight on the night before the holiday begins. Catholics have a similar spring-cleaning ritual leading up to Lent, either just prior to or within the first week of the forty-day Lenten period, which starts the day after Mardi Gras, or Fat Tuesday, in February.

Spring cleaning is especially common in environments that have cold winters. After a winter of houses being closed up, the warmer temperatures provide the opportunity to open doors and windows and air out the interior. Having grown up with a lot of Persian wool rugs in our house, our spring cleaning took place as soon as the weather turned warm, when we removed all of the rugs, beat them with a broom, and sat them in the sun for a day to kill all of the moth eggs.

In North America and Northern Europe, spring cleaning tends to occur in March or April and is a very therapeutic exercise in getting rid of things that are no longer of use. This is why spring is also the season for garage sales.

That said, you don't have to wait till spring. Studies show that cleaning is important to help us maintain our physical and mental health. Clearing out piles of clutter and cleansing thoroughly removes allergens such as dust mites, mold, mildew, and pet dander, along with droppings and saliva from pests like mice, cockroaches, and termites. If those aren't reasons enough to inspire cleaning at any time, I don't know what is.

LIKE NEW YEAR'S EVE, when many of us take the time to be introspective and make vows to improve ourselves, rituals of renewal can take many forms. Pressing the reset button on our personal health is one of them. People approach this in all kinds of ways, and leading up to my fiftieth birthday, I was excited when I decided to give myself a health overhaul. I just couldn't figure out how to make that happen. After much trial and error, my opportunity for renewal emerged from an unexpected source—my gut.

I decided to take on the challenge of a thirty-day gut detox. This meant no wine, no coffee, no gluten, no dairy, and no sugar for a month. Now, after my own journey to renew my gut, I am a firm believer in the mantra—*food is medicine*. As plenty of studies show, better food choices result in healthier minds and higher productivity.

In her book *Brain Food*, Lisa Mosconi says that foods like fatty fish and dark leafy greens, vegetable oils, complex carbohydrates, and berries help shape our cognitive health. Our brains are made up of the chemical compounds in the foods that we eat, which means, as Mosconi says, that whatever we eat will become our thoughts. Studies show how bad food choices—like french fries and milkshakes—can lead to depression and chronic disease, like diabetes. A recent study predicted that by 2030 half of all adults within the United States will suffer from obesity.

To complicate matters, another study showed that people have more body fat today than they did twenty years ago, despite eating and exercising the same amount. This is possibly due to environmental factors relating to large-scale farming and the use of antibiotics in livestock, which may have altered our gut biome. The overuse of antibiotics has been another challenge to our ability to maintain a healthy gut flora. Maintaining a healthy gut microbiome is similar to cultivating a healthy and diverse garden.

There has recently been a heightened awareness of how fragile and important our gut biota is, and not only for us, but for other species as well. Elephants already knew this and seed their own gut by eating the dung of other family members. Nature's fecal transplant of sorts.

There is increasing evidence that rising temperatures around the globe are impacting the symbiotic relationship that some species have with their gut biota. One study shows that prolonged drought causes a change in the makeup of grass communities, which in turn affects the diet of rodents, which impacts their gut microbiome in a negative way. Another study showed that a few degrees' increase in temperature in Kazakhstan in 2015 caused a flare-up of dangerous bacteria within the gut flora of the endangered saiga antelope, killing over two hundred thousand antelope across the range of their population.

In a mysterious die off of hundreds of elephants in Botswana in 2020, research points to a naturally occurring toxic bacterial that concentrated in stagnant water due to a drought. Another die off in Zimbabwe may have been caused by a natural respiratory bacteria that was involved in the Saiga antelope deaths. These are examples of how a bacterial imbalance can throw nature off kilter.

These examples highlight the importance of maintaining a healthy microbiome and show just how sensitive it is to perturbation. The good news is that we can take charge of our own gut flora with a renewal ritual and engage in a health reboot. I feel like I was gifted ten extra years by pressing the reset button on my overall health and immune system. Though it wasn't easy to get myself there.

Granted, I wasn't unhealthy. I just wanted to develop good habits through a reboot. We all know that if we don't get enough sleep, we can't function as well as if we had gotten in a solid eight hours of zzz's. The same is true for what we put in our body—junk in equals junk out. Eating healthy food keeps both body and mind fit.

The first four days of my gut detox were sheer hell. But five days in, I was amazed at how well I was sleeping and how much more energy I had. Despite how well I was doing, by the end of week two, I was already planning my "retox," if you will. I had a very simple list in mind: a bottle of red wine and a chocolate cake with butter cream frosting. Maybe not all at once, but that's what was on the menu. By day twenty-one, I actually no longer had the desire for a decadent binge.

I could feel the difference that my abstinence from acidic and inflammatory foods was having on my body. That was all the encouragement I needed to forge ahead.

Renewal is hard work, and our bodies aren't used to new regimes, despite how much better they can be for us. It wasn't about reaching a specific goal, it was about the journey—and most importantly, creating a lifelong ritual of renewal. That was the real goal.

What was extremely helpful was the support I got from doing the program as part of a group. Success doesn't have to depend solely on independent willpower; often we need a community that understands our struggles and will happily support our efforts. Being part of a group can offer transcendent strength that is hard to generate on one's own. And I also found that, as a collective, the good habits I was practicing were easier to ritualize.

Another tool that I found helpful during this process was meditation. Meditation in the morning before starting work helped empower me to be in control of my day. To get there, I had to unclutter my mind, just as I had uncluttered my gut and my closet.

Music was an important tool for me, particularly as a novice. To enter a meditative space, I played a recording of Indian flute and tabla music. It required my full attention to follow the music's meandering path, which went from slow and meditative to frenetic. Twenty minutes of listening allowed me to push out all other thoughts. Then, once my mind was clear, I turned the music off and tried to think of nothing for at least ten minutes. The simple idea that I could quiet my mind made me feel empowered. The singularity of focus helped with food decisions during the detox, as well as with work and life goals, and meditation remains part of my renewal rituals today.

The benefits of meditation are many. Researchers have shown that the act of mindful meditation fosters better brain health, including rewiring our brain to improve our ability to regulate our thoughts, behaviors, and emotions within as little as eleven hours of meditation over time. The science of meditation shows us how life-changing beneficial habits and daily personal rituals can be. It is inspiring to understand just how malleable our brains are.

Small changes in our habits can make a huge difference with regard to overall health and the richness of how we experience our lives. Taking care of ourselves by practicing rituals of renewal is the best place to start, in whatever form. This begins with getting enough rest, eating the right foods, and exercising.

Spring cleaning is not just about getting organized—it's simply one ritual of renewal that supports and improves our health. Renewal rituals can be seasonal cultural celebrations as well as day-to-day personal habits that help us manage our lives and well-being. Many nonhuman animals do this naturally. I will always be grateful for witnessing the humpback whale and her calf when they returned to their home that New Year's on Maui, as part of their migratory ritual of renewal. They helped spark my own inward journey.

10 · RITUALS OF TRAVEL & MIGRATION

FOREST BATHING AND A JOURNEY TO ONESELF

> "We need only travel enough to give our intellects an airing."
> —HENRY DAVID THOREAU

WAS LOOKING OUT THE BACK window of a Cessna 182 airplane, counting elephants in a landscape scorched by a recent fire. It was toward the end of the dry season in the Zambezi region of Namibia in 1994. Fires are prevalent during this time, and the clouds burgeoned with the promise of rain. The largest elephant migration still in existence in Africa was about to commence.

A group of fifty elephants was moving from the ashy teak forest, through the acacia open woodland, and toward the Kwando River for a late-afternoon drink. Another fifty were making their way through the riverine forest along the snaking river.

In the distance, several groups were just hitting the river's edge at Horseshoe—named for a U-shaped bend in the river. Two hundred elephants were already at this very popular place to gather and drink at sunset.

We were flying low and slow, parameters necessary for a census like this, which made the landscape full of elephants all the more dramatic. Two of us were doing the counting out either window, at the back of the aircraft, while my husband, Tim, navigated from the copilot seat. Tim was in charge of the census design, its implementation, and crunching the numbers.

When our census transect line ended at the river, the plane banked sharply to the right. We were headed south toward the border of Botswana to pick up the GPS coordinates of our next transect line.

As we flew down the river, we could see that the group of elephants at Horseshoe extended into group after group, each small family following head to rear along networks of dusty footpaths. These elephant migratory paths, centuries old, wove through the teak and open woodlands spanning hundreds of miles, looking like a giant network of collective memories.

I marveled at so many elephants filling the woodland from north to south. They spilled out from the forests, into the river, and on out to the horizon of the vast floodplain. The further we flew south, the denser the gray mass became, as more and more smaller elephant groups joined the larger extended herd. Eventually, we counted over one thousand elephants.

When the rainy season approaches in this region, elephant aggregations get larger and larger as they migrate south and into the Okavango Delta of Botswana. Until the rains begin, the elephants move in a three-day cycle, going from the river to inland in order to find enough food, and then back to the river again. Each time, they must go farther inland, as the pickings get slimmer. We were catching them at the river, just as they needed that big, long drink at the end of the third day.

The scene evoked a lost age from a century ago, before heavy poaching and before fences blocked animal migrations in many regions of Africa. However, the ancestral migration of elephants from the Zambezi to the Okavango Delta remains a true natural wonder.

Today, many of the traditional elephant migration routes are now blocked. In southern Africa, one once went from Etosha National Park, north into Angola, following the rain and new plant growth.

In the perennial river systems to the west, others headed north into Angola and Zambia and south into Botswana.

At one time, great wildebeest and zebra migrations occurred within Botswana, but most of those routes have been cut off by veterinary fences designed to keep wildlife separated from domestic cattle. For elephants in this region, the main route is from the Zambezi to the Okavango Delta in Botswana, which supports the largest free-ranging elephant population and migration route left in Africa. There are efforts to reconnect a safe passage between the southwest region of Zambia and the southeast region of Angola to the north and the Okavango Delta to the south through the Zambezi, which at one time was, and occasionally still is, an open migratory path for elephants.

In the Serengeti plains of Tanzania, over two million wildebeest and two hundred thousand zebras and small antelope migrate over eighteen hundred miles each year. They migrate to the Masai Mara in Kenya, seeking the epic grasslands that spring up with the fresh rain.

Animals migrate for ecological reasons in a complex dynamic that follows the pattern of the seasons. The snow goose, for example, migrates south to the United States and Mexico to escape cold winters in Greenland, Canada, and Alaska. If prey animals migrate, their predators follow, the same way humans once pursued herds of American buffalo or woolly mammoths. When rain arrives in a dry climate, triggering the rapid growth of plants, animals like elephants, zebras, and wildebeest walk as far as they must to reach them.

The eastern gray whale migrates from Russian waters to Mexico and back every year. They mate and give birth in the shallow, protected lagoons in Mexico, and then when the calves are strong enough, they return to the nutrient-rich, cold waters of the Arctic. Through satellite tracking, scientists recorded one gray whale traveling nearly fourteen thousand miles, the longest recorded migration by a mammal.

Of the 120 migratory bird species, the tiny arctic tern makes the longest migration of any animal in the world. Each year, it flies on the order of forty-four thousand miles, following zigzagging routes between Greenland and Antarctica.

Humans slowly shifted away from their hunter-gatherer and migratory behavior during the Pleistocene era, starting about twelve thousand years ago. This was likely due to a combination of factors: the warming climate as the Ice Age ended, the overhunting and extinction of major prey animals—like woolly mammoths, mastodons, and giant sloths—and the rise of domestication of livestock animals and agriculture.

Today, humans migrate mostly for socioeconomic or political reasons: to seek a better life or to escape oppression. Immigrants then must shape new identities and form new communities that offer a sense of belonging in their new environment. Many new rituals arise as a result, as people incorporate their previous beliefs and sense of self into their new lives. One example is the Ashura procession in Copenhagen. Every year, thousands of Muslims who have settled in Denmark from Iraq, Iran, and other Muslim countries join in a procession to celebrate their religion.

Some people travel as a part of a religious ritual, like Muslims traveling to the Saudi Arabian city of Mecca, the holy city of Islam. Roman Catholics pilgrimage to Rome to visit the Vatican, the home of the pope.

A pilgrimage is a type of journey, and many types of travel rituals exist today. For many British, Australian, and New Zealand students, taking a "gap year" to travel, both for self-discovery and to discover the world, is a common tradition, while many indigenous people also engage in travel rituals, both physical and metaphysical, to gain perspective on their lives. Indigenous Australians believe songlines, or "dream tracks," are paths across the land and sky created by ancestral spirits. These paths can be traversed through ceremony—by singing and listening to the ancient cultural songs, and listening to these songs is considered the same immersive experience as walking the actual songlines and seeing the land described in the songs. Some of these journeys span thousands of miles, traversing many different indigenous cultures.

Vision quests are another type of sacred ritual travel, and these vary in indigenous cultures and communities around the world. Typically, these involve fasting for several days and nights at a designated sacred site, in which the individual spends time in nature alone to gain wisdom. Traditionally, the vision quest is a rite-of-passage ritual for young men entering adulthood, but vision quests are now practiced by many adults of different cultural backgrounds seeking an intensive experience with nature, which sometimes includes hallucinogens. Questing involves elements of solitude, emptiness, vulnerability, and self-reliance.

Today, many people live very sedentary lives. Who needs to go anywhere when we can hold the world in our hands? However, while communications and digital technology might allow us to bring everything we need to our home, we still benefit from going out into the world. Travel, these days, is often considered a luxury, but many studies show how healthy it is, both physically and psychologically.

BACK IN 1994, NOT LONG BEFORE Tim and I began our last elephant count leading up to their migration, we unwittingly ended up driving through a large fire. We were checking on some tree transects to see how intensely elephants were eating their favorite species along the river during the dry season, when the wind picked up.

While I took measurements of seedling and sapling heights and the incidence of debarking of trees, hot ash swirled around my hiking boots. My socks and shins became completely covered in soot. We were moving fast and trying to finish all ten transects before our elephant census, but when one of my transect lines melted from the nearby fire, we knew it was time to get out of there.

During our retreat, the wind picked up further, and the fire suddenly crossed the road, both in front of and behind our truck. Then it quickly spread through the few remaining stumps of dry, yellow grass on either side of the already scorched road.

An elephant family crossed the sandy track in front of us, covered in black ash. Their skin was a sooty gray with thin, crepe-like wrinkles, as if they had been desiccated from a prolonged lack of water. Bleach-white ivory offset the deep gray of their skin. Despite the bleak environs, at least some patches of fresh grass had sprouted up for them to eat among the char.

I steered to the left of a series of white, ash-filled dust devils. These churned across the singed earth, behind the fire, collecting dry orange leaves that were still smoldering and steaming. We drove farther east, away from what looked like the freshest fire. The wind was blowing west, which meant that the thick smoke was billowing away from us.

Periodically, the fire got so hot that dead trees were spontaneously combusting and exploding. Voracious tongues of fire hungrily consumed the black wood, setting the horizon ablaze. One tree burst on the left and then another on the right, the explosions causing the air to get more and more turbulent, conjuring up more mini-twisters in their wake.

In the distance, the only trees standing were the giant leadwoods, peppering the landscape with shining, silvery-gray trunks, topped with bright orange leaves. Those with torched leaves would survive another year, while others stood as naked, erect carcasses, distinctive, individual characters of the habitat. Surprisingly, the twigs at the tops of such hardwood trees are so dense that they remained intact, and they probably would for many years to come.

As we got farther away from the fire, color returned to the landscape, the soil now clay instead of sandy. The papery pods of large terminalia trees gleamed purple, while pastel, neon-green; low grasses painted the middle of the hard-pan clay track. Around us, the open woodlands displayed a palette of orange, yellow, and rust.

Natural fires are a normal feature of many ecosystems. Some plant species need fire for their seeds to germinate. This is particularly true of chaparral ecosystems in California, where hot fires, every few decades, stimulate new plant growth. Many animals need fire to stimulate fresh plant growth, particularly at the end of the dry season in Africa, when there is very little to eat. Early humans developed a relationship with fire almost a millennium ago that shaped our evolution. Learning how to make fire gave us the confidence to take on new challenges and the unknowns involved with migrating from Africa into Eurasia.

Modern human activities have caused temperatures to rise, which has led to increase in the frequency and severity of fire within many ecosystems, including California's, that did not evolve to withstand this pressure—particularly the combination of fire, the grazing of domestic livestock, and the invasion of pests like bark beetles. This pressure affects both the habitat and all the animals that evolved to live in those environments.

Fire influences the migration patterns of many birds by altering natural plant communities. In some cases, fires benefit migratory species, and in others, they have a negative impact. A study of migratory birds in pine forests in Montana showed that populations of two species, the Cassin's vireo and the Swainson's thrush, declined after wildfires. Whereas the Townsend's solitaires and lazuli buntings increased. However, increases in some migratory birds were only found after moderate fire. With severe fires, all numbers decreased.

Tim and I were lucky that the fire we found ourselves driving through was not severe. We escaped the bush fire unscathed. Later, while in the airplane surveying the elephant migration, we got a different vantage of this burnt African savanna from the air. As we banked to avoid a group of vultures gliding in an updraft, we could see where the fire had been, still smoldering in places. Soon, the sun was getting low, so we returned to the reed-and-thatch house at Susuwe Ranger Station that had been our home for three years.

As we landed on a remote bush airstrip, the great, bloody orb of a sun sank behind us, silhouetting a giant baobab tree. As the sun slipped down to the horizon, it was erased by smoke, leaving us with a gray sky. A crimson full moon rose on the opposite horizon, reflected in the snaking river in the distance.

After a shower and a meal, we settled into bed, exhausted from the long day of flying. However, we were treated to mating hippos moaning throughout the night, which didn't make for the most restful sleep.

OUR JOURNEY IN AFRICA began in January of 1992. Tim and I had planned a gap year between degrees and arrived in South Africa with very little money. We bought his grandmother's VW Beetle and traveled between national parks in and around southern Africa for nine months before we were offered jobs in the Zambezi region of Namibia to study its elephant population.

At the beginning of the twentieth century, this region was carved out from the southern border of Zambia and the northern border of Botswana, above the Okavango Delta. This strip of land was the result of German colonists' efforts to gain access to the Zambezi River and make a trucking route across southern Africa before World War I.

When Tim and I arrived, the region had no fences separating wildlife from human settlements, which often led to competing interests. Elephants needed forests for food and access to the river to drink—and humans needed space for crops as well as safe access to the same water. Part of my job was to serve as an interface between the needs of both elephants and farmers and to come up with solutions to the inevitable conflicts that ensued.

I was always on tenterhooks, waiting for one of the game guards to come knocking on the door of our modest abode. He'd be exhausted from his many-mile bicycle ride along the hot, dusty track to report that the elephants had raided yet another farmer's cornfield. Having been hired by the government to help mitigate such problems, the ramifications of any elephant activity outside the reserves was viewed by the communities as my responsibility. It was impossible to patrol the entire floodplain, where human settlements lined the one side of the river and pristine elephant habitat on the other. With the weight of this impossible task on my shoulders, I never felt completely at rest while I lived in the Zambezi.

However, our reed-and-thatch house on the edge of the Kwando River was a source of great solace. In the tiny bedroom, knee-to-ceiling screen walls were the only barrier between us and outside—from the leopards patrolling at night and the elephants browsing in the early blue morning light. Elephants were very aware of our presence and gently stepped around our humble abode to access their favorite camelthorn tree pods.

At night, elephant breath on my face would wake me with a start, and I'd look up the nose of the world's largest land creature chewing on branches and pods just outside the window. The elephant's slow and methodical breathing would eventually lull me back to sleep.

When we left the Zambezi region in 1995 to return to graduate school, it was a tumultuous time. Angolan rebel forces were making incursions into the region, and our field station was under the threat of siege, less than ten miles from the border. In 2011 and 2012, more than fifteen years later, I eventually returned to conduct an elephant conservation study, which included the health of people living within the conservancies. I was curious to see how much had changed.

One thing was fences. Double fences, high enough that game like kudu couldn't jump over them, had been erected between Botswana and Namibia to minimize contact between wildlife and cattle in order for Botswana to sell disease-free beef to the European market. Yet these veterinary fences threatened one of the last elephant migration routes left in Africa, along with many other species that got caught up in the fencing. Tim's elephant movement data from the early 1990s was used to justify removing portions of the fencing next to the river. This made those early years of laborious radio-tracking of elephants through the bush—in the early days of satellite technology—very gratifying. At that time, GPS coordinates from satellite collar fixes were accurate within the range of a mile; today, tracking collars use GPS technology, making it possible to have accuracy within feet or less.

I had an ambitious schedule of community interviews in my old study region. One day, I took the opportunity for a break and found myself standing on the foundation where our home once stood, next to the Kwando River. I'm not sure what I expected to see when I returned. None of our colleagues had mentioned that our reed-and-thatch house no longer existed.

While staring at the few remaining bricks, I scrambled for a foothold on the home that we had built for ourselves. I thought the physical presence of the house would provide a conduit to the past. Without that, I was momentarily lost in memories that no longer felt real.

I breathed in a ragged breath and exhaled slowly, walking along the empty, twenty-foot cement slab, crunching through the dried sausage-tree leaves and large, dark red, velvety flowers. In the wet season, the fifteen-pound sausage-like fruits would come crashing down onto our flimsy thatched roof from twenty feet above.

The Zambezi region of Namibia might feel frozen in time, were it not for the shifting course of the river denoting the ticking of a primordial clock. A colleague had told me that a few years before, the largest rainfall in twenty years occurred, transforming this lowland into a wetland. Our house flooded out and the front yard became a backwater, which remained until the next drought.

I kicked a few leaves away from where the front door used to be. I stood where the mice rattled the makeshift front door just after dark, squeezing under and popping into the warm safety of our nest. Fortunately, the black mambas never got wise to the mother lode of small rodents that shared our home. All it would have taken was one quick bite to return us, prematurely, to the carbon cycle.

I walked past where my desk once sat, under which the scorpions scurried away from our feet on our way to the bathroom at night. Our homemade teak bar and counter had opened onto the reed-walled dining area. Tim spent many hours in that dining room, attempting to reassemble a baboon skeleton, and then the skeleton of a spotted eagle-owl we had found. Since the mud in the wet season made the roads unnavigable, we were forced to use our own minds for entertainment, and our minds were too restless to sit still.

Adjacent was the bathroom, where a spacious open shower once seemed like an incredible luxury. A few bricks still stood where we had to reinforce the shower wall. Every other brick that enclosed our modest living quarters had been anonymously claimed.

I turned to look at the floodplain and something immediately struck me. Although nature had taken back our African home, the old trees, towering above, were still exactly as I remembered them all those years ago.

The only tree that had fallen was the large camelthorn next to the sandy driveway, the one that the resident elephants would visit when the pods were ripe in July and August. The old bulls would press their foreheads up against the trunk of the tree and push just enough to set off a wave of vibrations. They knew exactly what they were doing. Shaking the tree like this released a shower of large, fuzzy, gray pods from the tree. From the looks of things, a combination of fire and elephant pod-seeking had done the tree in. It lay across the drive in a giant, twisted heap.

Looking up at the tree canopy, it was clear that it didn't matter whether there was a physical roof over my head. The trees made this place feel like home to me. My old friends stood familiar and open, like a gateway to my past. The expansive, gentle, *Acacia erioloba* limbs overhead had kept my memories safe all this time.

Immersed in reminiscing under heavy boughs that were levitating, I felt everything come into focus. The trees channeled my emotions, my memories, and ultimately, my consciousness, making me feel truly at one with the natural world.

The Japanese call this experience "forest bathing"—or *shinrin-yoku*, *shinrin* meaning "forest" and *yoku* meaning "bath." In Japan, forest bathing is a ritual that fosters a conscious awareness of the whole sensory experience of being in a forest. In a recent Japanese study, forest bathing was shown to reduce stress levels, lower pulse rate and blood pressure, and promote relaxation.

Seeing my old tree friends again sparked a flood of memories. Feelings of excitement about nature, about my life's adventure in Africa, feeling the roots I had long ago set down in this place—so many emotions poured out.

My 2011 return to Zambezi was fortuitous. Despite changes, this remained a relatively pristine and largely open environment, with few fences separating people from wildlife within the region. The unprotected area where we had once lived was now a national park.

I felt like I glimpsed a brighter future ahead, though not one without hurdles. One hurdle was the ongoing struggle with HIV, and another was the continued conflict between farmers and crop-raiding elephants. Yet thanks to strategic conservation policies and ecotourism, much of the land remained untouched. These policies allowed local communities to form conservancies that benefited from wildlife and the natural habitat, rather than increasing the number of farms to meet the demands of a rapidly growing human population.

I left the foundation of my former home and walked out onto the floodplain, where the locusts used to swarm in dark clouds during the still, dry years. Now, trees adorned with hanging weaver nests stood in the middle of a deep pan of water. Lechwe and reedbuck antelope dotted the edges of the bank. Hippos floated in the deep channel beyond, where we used to go at sunset, to be entertained by their open-mouthed threats, which displayed their formidable canines.

In 2019, I returned to Zambezi again to revisit the trees, the elephants, and old friends. Even as some wildlife conservation efforts continued to succeed, new challenges had developed and some old challenges remained. The one that may never go away is the incredible challenge of people living alongside wildlife in this open system, whose dangers are sometimes life-threatening. So, despite the benefits, the downsides need constant attention, as negative attitudes toward wildlife are a threat to any conservation efforts.

I was thrilled to learn the elephants were doing well in the region, but that meant the elephant-human interface was expanding. This makes it twice as hard for conservationists to develop strategies for sharing water, land, and food with minimal human-elephant conflict. Further, there was an urgent need to protect teak and rosewood forests from illegal harvesting. Corridors for elephant migrations were still intact, but were being threatened by competing development plans. More work was needed to secure these important ancestral migratory paths. International conservation groups have come together to address this problem and have penned a plan for a Kavango–Zambezi Transfrontier Conservation Area covering part of southeastern Angola, southern Zambia, Zimbabwe, the Zambezi region of Namibia, and Botswana.

The return to the region reminded me that there is always dynamic tension between the pressure to develop and the need to protect natural areas. Conservation requires constant vigilance and negotiation. And traveling helps us learn what the important conservation issues are around the world. It also helps us develop a stronger connection to the natural world.

PROBABLY THE MOST compelling way to feel a direct connection to nature is through an immersive wilderness experience. As a conservationist, I have been fortunate to have had many such experiences of forest bathing in Africa, but not as many within the United States, which is why I leapt at the chance to visit Yellowstone, one of America's wildlife gems, for the first time. The Greater Yellowstone Ecosystem is one of the largest nearly intact temperate ecosystems in the world. Traveling to Yellowstone National Park to see the charismatic megafauna and landscapes of the American West had long been on my bucket list. When a conference at Jackson Lake, Wyoming, brought me to the area, I was determined to fit in a quick excursion.

It was the last weekend of the fall season, when the park had hardly any visitors. The only downside was the unpredictable weather.

I drove to Yellowstone from Jackson Lake Lodge and got to my campsite late the first night in the bitter cold. The freezing air penetrated my thin nylon tent, my ears rang, and I barely slept.

Not being a morning person, I found it challenging to get up at five AM. But I had no choice if I was to get to the best location to see sunrise over the peaks of the Grand Tetons. A thermos filled with hot tea was a most helpful companion.

While I watched the morning sun bathe the cold, granite peaks in pink light, the unlikely high-pitched bugling of elk in rut echoed throughout the flats. The call of the wild rose with the mist above the frosted land and quelled any internal complaints over the early hour and my restlessness about being on my own on this trip.

A male moose slowly emerged from his bed, hidden within the tall grass. When he entered the willow flats and saw a challenging bull, he bleated and mooed at the intruder and urgently corralled his harem. A lot of energy is spent by males during the mating season to prevent their females from getting stolen.

The two males prepared for combat. Each held his head down as he lunged, threateningly, toward the other. They carried their big rack of antlers at clashing level, rocking them back and forth, before lurching forward and coming to blows.

Later in the morning, I drove through the Lamar Valley hoping to experience my first wolf encounter. The landscape lay open before me, immense and empty, with the red-orange walls of the valley rolling out in all directions, from the horizon to the sun. The scenery was so stunning that I almost didn't care whether I saw a wolf or not. Just being in the valley of the wolf was enough.

Then toward the bottom of the valley, I turned a corner and saw an old male wolf standing near the road. There was no sign of his pack. In awe, I slowed, stopped the car, and shut off the engine.

I knew this moment would be fleeting. He was mottled silver and steel gray on top, with mostly white legs and chest. His yellow eyes met mine for a few moments, his body completely still. I didn't dare look away to grab my camera.

The wolf took one last look with his head held low, sniffing toward me, then he turned and ran across the valley floor. As I watched him disappear into the forest in the distance, I imagined, for an instant, what it might be like to follow him on his wild itinerary.

Like the Zambezi in Namibia, the Yellowstone region is largely an open system with its own tensions related to the animal-human interface. Here, the conflicts aren't between corn farmers and elephants but between ranchers and wolves, the region's top predator. Many do not wish to have wolves as neighbors.

The gray wolf was reintroduced to Yellowstone in the mid-1990s, and since then, it has had a very positive impact on the environment. Wolves have helped regulate the elk population, and the reduction in elk has allowed willow and aspen to increase—after being overgrazed by elk—and that has fostered the recovery of many other species, including the return of beavers and increased songbird populations.

Later, on my way to my next campsite, I spotted a small group of bison grazing in the distance. I stopped at a turnout and walked up the trail to watch the herd cross the plain. I was aware of being in grizzly country, so I didn't stray far. Grizzly bear behavior is foreign to me, and they are notoriously more prone to aggression leading up to hibernation.

I arrived at Cascade Creek Falls just after sunset. The falls lie at the farthest reach of Mammoth Hot Springs, in the northern corner of Yellowstone National Park. A hydrothermal natural wonder, this strange phenomenon is a remnant of a volcanic eruption some six hundred thousand years ago.

As darkness fell, mist rose off the tiers of eerie green, travertine pools. Heat-loving bacteria looked like rust streaks running down a limestone layer cake. The cascading, white blankets of calcium carbonate were many thousands of years in the making.

The layered terraces of steaming ponds enveloped the bases of the surrounding fir trees. Sulfurous plumes crept along the surface of the pools, as if nature were brewing secret demons within these bubbling cauldrons.

I blew on my hands to warm them as a snowshoe hare hopped across my path along the wooden boardwalk, his ears half-cocked. He was not yet in his winter coat. It was now so cold that the warmth of the scalding, hot springs seemed particularly inviting. Unfortunately, I was too exhausted to make a side trip to the only spring safe for a hot soak. Perhaps next time.

My short trip to Yellowstone, despite its occasional discomforts, worked its magic. I felt revived and renewed. I am not the only one to experience this, of course. Journeys in nature, and even travel itself, are restorative. The travel bug is encoded in our genes. Our ancestors were migratory hunter-gatherers, and their migrations eventually encompassed the globe. Ten thousand years or so of a more sedentary, agrarian, and now suburban and urban lifestyle haven't fundamentally changed habits formed over millennia. We like to be outside and on the move.

All sorts of studies back this up. They confirm that even planning a trip makes us happier. The anticipation of travel is a stress release, and people who take regular vacations are less likely to have heart disease or heart attacks. A trip can improve blood pressure and boost our immune system. Experiencing new surroundings offers us a fresh perspective on life, as well as a new appreciation for the place we call home, all of which cause dopamine levels to increase in the same way that a new love does—which explains why travel can be so exhilarating. These benefits happen both during the trip and later when reminiscing.

The experiences of travel stimulate the brain to develop new neurons, which contributes to more creative thinking and novel ideas. A study of students who studied abroad found that they were 20 percent more likely (than those who didn't study abroad) to succeed at problem-solving tasks. These studies confirm what we discover ourselves directly when we leave the house, travel somewhere new, meet new people, and immerse ourselves in our stunning natural world.

IN 2019, ON ONE of my last days in Zambezi, I had the opportunity to sit at my favorite place to watch elephants coming to drink at sunset: at Horseshoe on the Kwando River, the same place I had flown over during our elephant censuses in the early 1990s. As I watched a group of thirsty elephants rush down the sandy bank to the river's edge, I felt a sense of hope despite all the challenges facing the region.

My reasons for hope were right in front of me—the habitat looked exactly as it had when I first worked here almost thirty years ago. And there were more elephants. As more and more elephants poured into the river and engaged in their familiar greeting rituals,

I felt privileged to witness these reunions, which have been going on for thousands of years. Today, despite modern human impacts, the elephants in this region are surviving and even thriving, with their recent numbers rising higher than we had ever recorded.

It's very gratifying to me that elephant populations are increasing in the Zambezi region, at a time when poaching is still decimating elephant populations elsewhere. It makes me hopeful for the elephant's future.

In 2019, I had another hopeful and satisfying "last day" experience on Maui, where Tim and I had lived during spring and fall for a period of five years. We were packing up our lives there and switching our spring and fall migration to Boston. On our last day, we went snorkeling in Wailea to say goodbye to our favorite swimming spot. It was March, and the humpbacks were in Hawaii to mate and calve again.

With our heads submerged in the water, we could hear what sounded like a mother and calf calling back and forth, as I followed four hawksbill turtles over a live coral bed. In the late 1980s, when I was in graduate school in Hawaii, humpbacks and hawksbills were endangered, and now they were both in abundance here. I felt another moment of hope for our stewardship of this planet and its creatures.

We are connected to elephants, whales, wolves, and all other sentient beings in ways that may not seem obvious, although it's perhaps more obvious than the fact that we share 50 percent of our genes with the banana. We have the power to protect or destroy our surrounding habitat and all the other citizens that share this extremely unique planet with us. This is particularly important as the impacts of climate change increase—from devastating hurricanes, floods, and wildfires to diseases and coral reef die-offs. Whether disasters are natural or unnatural, everyone is affected. And if we make a conscious decision to save other species and habitats, we also save ourselves.

Which is why the ten rituals in this book are so essential. They represent all the ways we've learned to build a stronger self, stronger relationships, and a stronger connection to the world.

As for forest bathing, or immersion in nature, opportunities exist all around us. We don't need to trek the Appalachian Trail or climb the Rockies or Sierras. Camp for a weekend in a local state park. Go for a walk, hike, bike ride, or paddle where you live. Even an hour or two in nature will recharge us and give us a renewed sense of priorities.

Take longer trips when you can. Unplug when you travel as much as possible. The benefits of nature immersion and travel come as much from how we focus our attention—awakening to the world and others as well as to the world within—as where we go.

This awakening can happen in our own yard. As I write these words, I am sitting in my backyard. It's the middle of spring and the bright yellow-and-black orioles are stripping the edges of the fronds of the fan palms for nesting material. The female flies off with a long string of frond dangling from her beak, as the male follows behind. A newly emerged monarch butterfly tests its wings for the first time, as a black phoebe eyes it, using my hammock as a perch.

A male Anna's hummingbird dive-bombs a competitor as a pair of house finches build a nest, deep inside the cypress that we planted in our yard almost fifteen years ago. As the female stitches her nest together, the male chortles away on top of the tree. Most days, I hear them twittering in the early morning, right outside our bedroom window. A welcome change to the northern mockingbird that keeps me up much of the night in the spring. Although I marvel over their repertoire of songs, I wish they didn't have so much energy in the middle of the night.

In *Everything in Its Place*, Oliver Sacks writes: "I cannot say exactly how nature exerts its calming and organizing effects on our brains, but I have seen in my patients the restorative and healing powers of nature and gardens, even for those who are deeply disabled neurologically. In many cases, gardens and nature are more powerful than any medication."

With every detail I notice about nature unfolding in my backyard, I become more introspective. I am living in the moment, rather than thinking about the future, as I often do. Sometimes we spend so much time working toward a goal that we end up missing the life we are living. Life promises to be a lot more rewarding if we enjoy each moment as it occurs, rather than constantly anticipating and worrying about what life will be like in the future. I have realized over the years that the farther I get away from home, the more beautiful it becomes. And that makes me appreciate home all the more when I'm there. But I do have to pinch myself sometimes as it is easy to get distracted and become too busy to pay attention to the little things—the most important things.

My husband and I have had our home base in San Diego for the past fifteen years, and we have planted every flower, shrub, tree, and vegetable in our backyard and along our terraced bank. As I breathe in the scent of pine and the coolness of evening, I appreciate all the hard work we've put into this garden. I feel a deep attachment to the trees we planted. They have become part of us as we have become a part of them, cycling carbon between plant, ground, air, and animal. It's no wonder that gardening reduces stress and is correlated with increased longevity.

Immersing ourselves in nature is a journey, one that leads to better health and increased compassion, in the Buddhist sense of the word. In Buddhism, compassion is the result of knowing one is part of a greater whole and is interdependent and connected to that whole. Our rituals help us to not lose sight of ourselves and our place in the natural world.

When we have a strong sense of self, it empowers us to mindfully greet one another, to gain strength from the group, to court those we love, to gift a stranger with kindness, to sing out in joy, to hold hands and gaze into one another's eyes, to play silly games, to commemorate a lost loved one, and to renew a commitment to our mind and body. Nature is the gateway to reengaging in our wild rituals for a richer, more rewarding life.

NOTES

INTRODUCTION

page 4, Genes we share with the banana: Hoyt, A. 2019. "Do People and Bananas Really Share 50 Percent of the Same DNA?" How Stuff Works. https://science.howstuffworks.com/life/genetic/people-bananas-share-dna.htm.

page 4, Recent genetic findings point to all current life on Earth: Weiss, M. C., F. L. Sousa, N. Mrnjavac, S. Neukirchen, M. Roettger, S. Nelson-Sathi, and W. F. Martin. 2016. "The Physiology and Habitat of the Last Universal Common Ancestor." *Nature Microbiology* 1 (9): 16116.

page 4, Single-celled organisms needed three billion years: Pennisi, E. 2018. "The Momentous Transition to Multicellular Life May Not Have Been So Hard After All." *Science Magazine*. https://www.sciencemag.org/news/2018/06/momentous-transition-multicellular-life-may-not-have-been-so-hard-after-all.

pages 4–5, All multicellular organisms (including humans and bananas) share: ibid.

page 5, Gill slits are found in the embryos of all vertebrates: Graham, A., and J. Richardson. 2012. "Developmental and Evolutionary Origins of the Pharyngeal Apparatus." *BMC EvoDevo* 3 (24): 1–8.

page 5, Eight million years ago, mammals as diverse as horses: O'Leary, M. A., J. I. Bloch, J. J. Flynn, T. J. Gaudin, A. Giallombardo, N. P. Giannini, S. L. Goldberg, et al. 2013. "The Placental Mammal Ancestor and the Post–K-Pg Radiation of Placentals." *Science* 339 (6120): 662–667.

page 5, Modern brain-imaging technology: Adolphs, R. 2009. "The Social Brain: Neural Basis of Social Knowledge." *Annual Review of Psychology* 60: 693–716.

page 5, The same hormones are expressed: Soares, M. C., R. Bshary, L. Fusani, W. Goymann, M. Hau, K. Hirschenhauser, and R. F. Oliveira. 2010. "Hormonal Mechanisms of Cooperative Behaviour." *Philosophical Transactions of the Royal Society of London, Series B, Biological Sciences* 365 (1553): 2737–2750.

page 5, Many animals also experience a lot of the same emotions we do: Boissy, A., and H. W. Erhard. 2014. "How Studying Interactions between Animal Emotions, Cognition, and Personality Can Contribute to Improve Farm Animal Welfare." In *Genetics and the Behavior of Domestic Animals*, edited by T. Grandin and M. J. Deesing, 81–113. London: Elsevier.

page 6, Consider a lone male chimpanzee approaching a large fig tree: Kuhl, H. S., A. K. Kalan, M. Arandjelovic, F. Aubert, L. D'Auvergne, A. Goedmakers, S. Jones, et al. 2016. "Chimpanzee Accumulative Stone Throwing." *Scientific Reports* 6: 22219.

page 6, Ritual is called "accumulative rock throwing": ScienceVio. 2016. "Chimpanzee Accumulative Stone Throwing." https://www.youtube.com/watch?v=eVv3IUGPDK8.

page 7, The selected trees had either buttresses or hollow trunks: Kalan, A. K., E. Carmignani, R. Kronland-Martinet, S. Ystad, J. Chatron, and M. Aramaki. 2019. "Chimpanzees Use Tree Species with a Resonant Timbre for Accumulative Stone Throwing." *Biology Letters* 15 (12): 20190747.

page 7, These patterns could even be a precursor to musical rhythm: Weiler, N. 2015. "Chimpanzees Drum with Signature Style." *Science Magazine*. https://www.sciencemag.org/news/2015/01/chimpanzees-drum-signature-style.

page 7, Chimpanzee rituals are thought to be homologous to our own: Hopper, L. M., and S. F. Brosnan. 2012. "Primate Cognition." *Nature Education Knowledge* 5 (8): 3.

page 7, Chimpanzees enact a ritual dance at the onset of rain: Harrod, J. B. 2014. "The Case for Chimpanzee Religion." *Journal for the Study of Religion, Nature, and Culture* 8 (1): 8–45.

page 7, Jane Goodall suggests that chimpanzees: Goodall, J. 2005. "Do Chimpanzees Have Souls?" In *Spiritual Information: 100 Perspectives on Science and Religion*, edited by C. L. Harper Jr. and J. Templeton, 602. Philadelphia: Templeton Foundation Press.

page 8, Stone-accumulation shrines at sacred trees: Kuhl et al., "Chimpanzee Accumulative Stone Throwing," 2016.

page 8, Archaeologists recently found the oldest fossil evidence of a sacred ritual site: Coulson, S., S. Staurset, and N. Walker. 2011. "Ritualized Behavior in the Middle Stone Age: Evidence from Rhino Cave, Tsodilo Hills, Botswana." *PaleoAnthropology* 2011: 18–61.

page 8, A ritual is a specific act or series of acts: Hobson, N. M., J. Schroeder, J. L. Risen, D. Xygalatas, and M. Inzlicht. 2018. "The Psychology of Rituals: An Integrative Review and Process-Based Framework." *Personality and Social Psychology Review* 22 (3): 260–284.

page 9, Tennie and van Schaik, offer a very narrow definition of ritual: Tennie, C., and C. P. van Schaik. 2020. "Spontaneous (Minimal) Ritual in Non-human Great Apes?" *Philosophical Transactions of the Royal Society B* 375 (1805): 20190423.

page 9, Engaging in ritual can relieve stress: Eilam, D., R. Zor, H. Szechtman, and H. Hermesh. 2006. "Rituals, Stereotypy and Compulsive Behavior in Animals and Humans." *Neuroscience and Biobehavioral Reviews* 30 (4): 456–471.

page 9, Decrease anxiety: Brooks, A. W., J. Schroeder, J. L. Risen, F. Gino, A. D. Galinsky, M. I. Norton, and M. E. Schweitzer. 2016. "Don't Stop Believing: Rituals Improve Performance by Decreasing Anxiety." *Organizational Behavior and Human Decision Processes* 137: 71–85.

page 9, Make us more present: Marshall, D. A. 2002. "Behavior, Belonging, and Belief: A Theory of Ritual Practice." *Sociological Theory* 20 (3): 360–380.

page 9, Improve our congition: Rossano, M. J. 2009. "Ritual Behaviour and the Origins of Modern Cognition." *Cambridge Archaeological Journal* 19 (2): 243–256.

page 9, When we exaggerate a familiar behavior: Smith, A. C. T., and B. Stewart. 2011. "Organizational Rituals: Features, Functions and Mechanisms." *International Journal of Management Reviews* 13 (2): 113–133.

page 9, Improve our concentration: Tian, A. D., J. Schroeder, G. Häubl, J. L. Risen, M. I. Norton, and F. Gino. 2018. "Enacting Rituals to Improve Self-Control." *Journal of Personality and Social Psychology* 114 (6): 851–876.

notes

page 10, Tool to communicate and express intentions: Marshall, "Behavior, Belonging, and Belief," 2002.

page 10, They also create a mutual language to facilitate connection: Norton, M. I., and F. Gino. 2014. "Rituals Alleviate Grieving for Loved Ones, Lovers, and Lotteries." *Journal of Experimental Psychology: General* 143 (1): 266–272.

page 10, Rituals originating in early human societies: Renfrew, C., I. Morley, and M. Boyd. 2017. *Ritual, Play and Belief, in Evolution and Early Human Societies.* Cambridge, UK: Cambridge University Press.

page 10, Engaging in ritual is appeasing and mitigates anxieties: Smith, A. C. T., and B. Stewart. 2011. "Organizational Rituals: Features, Functions and Mechanisms." *International Journal of Management Reviews* 13 (2): 113–133.

page 10, During the performance of a group ritual: Norton and Gino, "Rituals Alleviate Grieving for Loved Ones, Lovers, and Lotteries," 2014.

page 10, Isolation is a major risk factor: Steptoe, A., A. Shankar, P. Demakakos, and J. Wardle. 2013. "Social Isolation, Loneliness, and All-Cause Mortality in Older Men and Women." *Proceedings of the Natural Academy of Sciences USA* 110 (15): 5797–5801. See also Holwerda, T. J., A. T. Beekman, D. J. Deeg, M. L. Stek, T. G. van Tilburg, P. J. Visser, B. Schmand, C. Jonker, and R. A. Schoevers. 2012. "Increased Risk of Mortality Associated with Social Isolation in Older Men: Only When Feeling Lonely? Results from the Amsterdam Study of the Elderly (AMSTEL)." *Psychological Medicine* 42 (4): 843–853.

page 11, Ritual plays an important role in bringing us together: Hobson et al., "The Psychology of Rituals," 2018.

page 11, This trunk-to-mouth ritual is akin to a handshake and signals respect: O'Connell, C. 2015. *Elephant Don: The Politics of a Pachyderm Posse.* Chicago: University of Chicago Press.

page 11, Modern society has made it easy for us to deprioritize rituals: Bone, A. 2016. "Why Rituals Are Still Relevant." Special Broadcasting Service. https://www.sbs.com.au/topics/voices/culture/article/2016/06/27/why-rituals-are-still-relevant.

page 11, You feel about social media: Ormerod, K. 2018. *Why Social Media Is Ruining Your Life.* London, UK: Cassell.

page 11, There's a reason why solitary confinement: House, J. S., K. R. Landis, and D. Umberson. 1988. "Social Relationships and Health." *Science* 241 (4865): 540–545.

page 11, Our most natural state of social interaction is touch or physical proximity: Chillot, R. 2019. "The Power of Touch." *Psychology Today.* https://www.psychologytoday.com/us/articles/201303/the-power-touch.

page 11, Lack of connection to a community: Ozbay, F., D. C. Johnson, E. Dimoulas, C. A. Morgan, D. Charney, and S. Southwick. 2007. "Social Support and Resilience to Stress: From Neurobiology to Clinical Practice." *Psychiatry (Edgmont)* 5 (35): 35–40.

page 12, Stress-related diseases are at an all-time high: "Stress Is a Leading Cause of Premature Deaths." 2019. The American Institute of Stress. https://www.stress.org/stress-is-a-leading-cause-of-premature-deaths.

page 12, Social media and technology are a double-edged sword: Ormerod, *Why Social Media Is Ruining Your Life.*

page 12, The trauma people experienced through physical isolation will be transgenerational: Bogin, B., and C. Varea. 2020. "COVID-19, Crisis, and Emotional Stress: A Biocultural Perspective of Their Impact on Growth and Development for the Next Generation." *American Journal of Human Biology*: e23474.

page 13, Mouth-licking in wolves: Mech, L. D. 1999. "Alpha Status, Dominance, and Division of Labor in Wolf Packs." *Canadian Journal of Zoology* 77 (8): 1196–1203.

page 13, Handshake in humans, are a form of information gathering: Huwer, J. 2003. "Understanding Handshaking: The Result of Contextual, Interpersonal and Social Demands." Psychology Department, Haverford College. http://hdl.handle.net/10066/757.

page 13, Evolved among social animals to strengthen bonds and build trust: Witham, J. C., and D. Maestripieri. 2003. "Primate Rituals: The Function of Greetings between Male Guinea Baboons." *Ethology* 109: 847–859.

page 13, Battle cries before war or sporting events, generate a sense of unity and common purpose: Wiltermuth, S. S., and C. Heath. 2009. "Synchrony and Cooperation." *Psychological Science* 20 (1): 1–5.

page 13, Roars of howler monkeys defending their turf at dawn and dusk: da Cunha, R. G. T., and E. Jalles-Filho. 2007. "The Roaring of Southern Brown Howler Monkeys (*Alouatta guariba clamitans*) as a Mechanism of Active Defence of Borders." *Folia Primatologica* 78: 259–271.

page 13, Territorial roars of a pride of lions: Grinnell, J., and K. McComb. 2001. "Roaring and Social Communication in African Lions: The Limitations Imposed by Listeners." *Animal Behaviour* 62 (1): 93–98.

page 13, Vocal rituals can also be a tool to release aggression and create boundaries: Briefer, E. F. 2012. "Vocal Expression of Emotions in Mammals: Mechanisms of Production and Evidence." *Journal of Zoology* 288: 1–20.

page 13, Elephant family's coordinated calls while departing from a waterhole: O'Connell-Rodwell, C. E., J. D. Wood, M. Wyman, S. Redfield, S. Puria, and L. A. Hart. 2012. "Antiphonal Vocal Bouts Associated with Departures in Free-Ranging African Elephant Family Groups (*Loxodonta africana*)." *Bioacoustics* 21 (3): 215–224.

page 13, Smiling and laughing, have been around for more than five million years: Davila-Ross, M., G. Jesus, J. Osborne, and K. A. Bard. 2015. "Chimpanzees (*Pan troglodytes*) Produce the Same Types of 'Laugh Faces' When They Emit Laughter and When They Are Silent." *PLOS ONE* 10 (6): e0127337.

page 14, Gazing into another's eyes is a powerful ritual of courtship and other forms of bonding: Fishbane, M. D. 2015. "Cultivating Connection: Reviving the Lost Art of Eye Contact." Good Therapy. https://www.goodtherapy.org/blog/cultivating-connection-reviving-the-lost-art-of-eye-contact-0527155.

page 14, Play affords individuals the opportunity to experiment with their surroundings: Burghardt, G. M. 2005. "Chapter 1: The Nature of Play." In *The Genesis of Animal Play: Testing the Limits*. Cambridge, MA: MIT Press.

page 14, Lion cub practicing hunting skills by pretending a littermate: Lancy, D. F. 1980. "Play in Species Adaptation." *Annual Review of Anthropology* 9 (1): 471–495.

page 14, Fantastical world in a sandbox, which fosters coping skills: Nijhof, S. L., C. H. Vinkers, S. M. van Geelen, S. N. Duijff, E. J. M. Achterberg, J. van der Net, R. C. Veltkamp, et al. 2018. "Healthy Play, Better Coping: The Importance of Play for the Development of Children in Health and Disease." *Neuroscience and Biobehavioral Reviews* 95: 421–429.

page 14, Rituals of grieving and healing from the loss of a loved one, including carrying or burying: Goldman, J. G. 2012. "Death Rituals in the Animal Kingdom." BBC Future. https://www.bbc.com/future/article/20120919-respect-the-dead.

GREETING RITUALS

page 18, Reinforce bonds between two close associates: Matoba, T., N. Kutsukake, and T. Hasegawa. 2013. "Head Rubbing and Licking Reinforce Social Bonds in a Group of Captive African Lions, *Panthera leo*." *PLOS ONE* 8 (9): e73044.

page 18, Reduce tension and foster reconciliation: Dal Pesco, F., and J. Fischer. 2018. "Greetings in Male Guinea Baboons and the Function of Rituals in Complex Social Groups." *Journal of Human Evolution* 125: 87–98.

page 18, Signal submission to a dominant individual: de Waal, F. B. M. 1986. "The Integration of Dominance and Social Bonding in Primates." *Quarterly Review of Biology* 61 (4): 459–479.

page 19, Both male and female hyenas present their erect genitalia: Smith, J. E., K. S. Powning, S. E. Dawes, J. R. Estrada, A. L. Hopper, S. L. Piotrowski, and K. E. Holekamp. 2011. "Greetings Promote Cooperation and Reinforce Social Bonds among Spotted Hyaenas." *Animal Behaviour* 81 (2): 401–415.

page 19, They improve existing relationships and help forge new ones: Rossano, M. J. 2009. "Ritual Behaviour and the Origins of Modern Cognition." *Cambridge Archaeological Journal* 19 (2): 243.

page 22, Often a greeting expresses different levels of respect: Lundmark, T. 2009. *Tales of Hi and Bye: Greeting and Parting Rituals around the World*. Cambridge, UK: Cambridge University Press.

page 23, A submissive greeting allows one to acknowledge their place in the social pecking order: de Waal, "The Integration of Dominance and Social Bonding in Primates," 1986.

page 23, Eight-step greeting ritual for British royalty: "Greeting a Member of the Royal Family." British Royal Family. https://www.royal.uk/greeting-member-royal-family.

page 23, For male elephants, the purposeful act of placing a trunk in a dominant male's mouth: O'Connell, *Elephant Don*, 2015.

page 23, The gorilla or chimpanzee hug: *The Routledge Handbook of Philosophy of Animal Minds*. 2017. Edited by K. Andrews and J. Beck. London: Routledge.

page 23, The bonobo kiss: Dias, P. A. D., and A. Rangel-Negrín. 2017. "Affiliative Contacts and Greetings." In *The International Encyclopedia of Primatology*, edited by A. Fuentes. Hoboken, NJ: John Wiley & Sons.

page 23, Zebra nip: McDonnell, S. M., and A. Poulin. 2002. "Equid Play Ethogram." *Applied Animal Behaviour Science* 78 (2–4): 263–290.

page 26, Their displays are similar to wolves: Siniscalchi, M., S. D'Ingeo, M. Minunno, and A. Quaranta. 2018. "Communication in Dogs." *Animals* 8 (8): 131.

page 27, Exchange hormonal information in a handshake: Frumin, I., O. Perl, Y. Endevelt-Shapira, A. Eisen, N. Eshel, I. Heller, M. Shemesh, et al. 2015. "A Social Chemosignaling Function for Human Handshaking." *eLife* 4: e05154.

page 27, The human handshake evolved for other reasons as well: "Handshake History." 2015. Deep English. https://deepenglish.com/2014/07/handshake-history-listening-fluency-116/.

page 28, Quakers adopted the handshake greeting: Corfield, P. J. 2017. "From Hat Honour to the Handshake: Changing Styles of Communication in the Eighteenth Century." In *Hats Off, Gentlemen! Changing Arts of Communication in the Eighteenth Century*, edited by P. J. Corfield and L. Hannan. Paris: Honoré Champion.

page 28, The modern French kiss on the cheek, la bise: "La Bise: The Art of French Kissing!" 2020. Insidr. https://insidr.co/la-bise-the-art-of-french-kissing/.

page 28, Scientists have found that when a stranger returns a smile: Lickerman, A. 2012. "Smiling at Strangers: How the Simplest of Gestures Can Spread Joy for Years." *Psychology Today*. https://www.psychologytoday.com/us/blog/happiness-in-world/201202/smiling-strangers.

page 29, People who engage within community: Umberson, D., and J. K. Montez. 2010. "Social Relationships and Health: A Flashpoint for Health Policy." *Journal of Health and Social Behavior* 51 (S): S54–S66.

page 29, The importance of physical and proximal social experiences: Short, S. E., and S. Mollborn. 2015. "Social Determinants and Health Behaviors: Conceptual Frames and Empirical Advances." *Current Opinion in Psychology* 5: 78–84.

GROUP RITUALS

page 32, Tarpon group hunt: O'Neill, M. P. 2018. "Mullet Mania." *Hakai Magazine*. https://www.hakaimagazine.com/videos-visuals/mullet-mania/.

page 33, How humpback whales hunt in Alaska, using a technique called "bubble-net feeding": Kosma, M. M., A. J. Werth, A. R. Szabo, and J. M. Straley. 2019. "Pectoral Herding: An Innovative Tactic for Humpback Whale Foraging." *Royal Society Open Science* 6 (10): 191104.

page 33, Dolphins do something similar: Gazda, S. J., R. C. Connor, R. K. Edgar, and F. Cox. 2005. "A Division of Labour with Role Specialization in Group-Hunting Bottlenose Dolphins (*Tursiops truncatus*) off Cedar Key, Florida." *Proceedings of the Royal Society B: Biological Sciences* 272: 135–140.

page 34, Separated from sardines by tens of millions of years of evolution, anchovies: Checkley, D. M., Jr., R. G. Asch, and R. R. Rykaczewski. 2017. "Climate, Anchovy, and Sardine." *Annual Review of Marine Science* 9: 469–493.

page 34, Sailfish counter this strategy by attempting to isolate a group of fish: Clua, E., and F. Grosvalet. 2011. "Mixed-Species Feeding Aggregation of Dolphins, Large Tunas and Seabirds in the Azores." *Aquatic Living Resources*. 14 (1): 11–18.

page 35, Each female has a specific position in the hunt that she always assumes: Stander, P. E. 1992. "Cooperative Hunting in Lions: The Role of the Individual." *Behavioral Ecology and Sociobiology* 29 (6): 445–454.

page 36, Chimpanzees hunting the colobus monkey: Watts, D. P., and J. C. Mitani. 2002. "Hunting Behavior of Chimpanzees at Ngogo, Kibale National Park, Uganda." *International Journal of Primatology* 23 (1): 1–28.

page 36, Honeyguide bird and honey badger: Fincham, J. E., R. Peek, and M. B. Markus. 2017. "The Greater Honeyguide: Reciprocal Signaling and Innate Recognition of a Honey Badger." *Biodiversity Observations* 8 (12): 1–6.

page 36, Honeyguide and different groups of African people: Spottiswoode, C. N., K. S. Begg, and C. M. Begg. 2016. "Reciprocal Signaling in Honeyguide-Human Mutualism." *Science* 353 (6297): 387–389.

page 36, Cooperative-hunting behavior in forest chimps is thought to provide clues to the evolution: Boesch, C., and H. Boesch. 1989. "Hunting Behavior of Wild Chimpanzees in the Tai National Park." *American Journal of Physical Anthropology* 78: 547–573.

page 36, Group rituals may have evolved in human societies from the need for a group of men to cooperate on a hunt: ibid.

page 36, The San people of southern Africa: "The San." 2019. South African History Online. https://www.sahistory.org.za/article/san.

page 36, Inuit of the Arctic and subarctic regions of Siberia, North America, and Greenland: "The Inuit." Facing History and Ourselves. https://www.facinghistory.org/stolen-lives-indigenous-peoples-canada-and-indian-residential-schools/historical-background/inuit.

page 36, Mayan peasant-hunters across the Yucatán Peninsula in Mexico: Santos-Fita, D., E. J. Naranjo, E. I. Estrada, R. Mariaca, and E. Bello. 2015. "Symbolism and Ritual Practices Related to Hunting in Maya Communities from Central Quintana Roo, Mexico." *Journal of Ethnobiology and Ethnomedicine* 11: 71.

page 36, The traditional Inuit believed that animals were superior to them: King, D. C. 2008. "Chapter 3: Inuit Beliefs." In *The Inuit*. New York: Marshall Cavendish Benchmark.

page 37, The Loojil Ts'oon, or Carbine Ceremony, of the Mayan peasant-hunters: Santos-Fita et al. "Symbolism and Ritual Practices Related to Hunting in Maya Communities from Central Quintana Roo, Mexico," 2015.

page 37, Define territories and boundaries: Eilam et al., "Rituals, Stereotypy and Compulsive Behavior in Animals and Humans," 2006.

page 37, Prepare for battle: Legare, C. H., and R. E. Watson-Jones. 2015. "The Evolution and Ontogeny of Ritual." In *The Handbook of Evolutionary Psychology*, edited by D. Buss, 829–847. Hoboken, NJ: John Wiley & Sons.

page 37, Communicate over long distances: Kuhl et al., "Chimpanzee Accumulative Stone Throwing," 2016.

page 37, Used to attract mates: Alcorta, C. S., and R. Sosis. 2007. "Culture, Religion, and Belief Systems." In *Encyclopedia of Human-Animal Relationships: A Global Exploration of Our Connections with Animals*, edited by M. Beckoff, 559–605. Westport, CT: Greenwood Press.

page 37, Trigger courtship: ibid.

page 37, Galvanize a group for a cause: Strandburg-Peshkin, A., D. R. Farine, I. D. Couzin, and M. C. Crofoot. 2015. "Shared Decision-Making Drives Collective Movement in Wild Baboons." *Science* 348 (6241): 1358–1361.

page 37, Engender trust and group identity: Watson-Jones, R. E., and C. H. Legare. 2016. "The Social Functions of Group Rituals." *Current Directions in Psychological Science* 25 (1): 42–46.

page 37, Provide identity to individual groups that were increasingly composed of nonkin: Hill, K. R., B. M. Wood, J. Baggio, A. M. Hurtado, and R. T. Boyd. 2014. "Hunter-Gatherer Interband Interaction Rates: Implications for Cumulative Culture." *PLOS ONE* 9 (7): e102806.

page 38, Evolution of a larger brain, cultural complexity, and language in humans: Muthukrishna, M., M. Doebeli, M. Chudek, and J. Henrich. 2018. "The Cultural Brain Hypothesis: How Culture Drives Brain Expansion, Sociality, and Life History." *PLOS Computational Biology* 14 (11): e1006504.

page 38, Our ancestors' brains tripled in size: Schoenemann, P. T. 2006. "Evolution of the Size and Functional Areas of the Human Brain." *Annual Review of Anthropology* 35: 379–406.

page 38, Brains had to expand to store and manage greater amounts of information: Muthukrishna et al., "The Cultural Brain Hypothesis," 2018.

page 38, Advent of language: Bundy, W. M. 2007. "Chapter 13: Models and Chemistry of the Modern Mind." In *Out of Chaos: Evolution from the Big Bang to Human Intellect.* Boca Raton, FL: Universal Publishers.

page 38, Also gain the loyalty of others by participating in a common ritual: Watson-Jones and Legare, "The Social Functions of Group Rituals," 2016.

page 38, Coordinated caring of offspring: Silk, J. B., S. C. Alberts, and J. Altmann. 2003. "Social Bonds of Female Baboons Enhance Infant Survival." *Science* 302 (5648): 1231–1234.

page 38, Leveraging group knowledge maximizes the likelihood of survival: Whitehouse, H., and J. A. Lanman. 2014. "The Ties That Bind Us." *Current Anthropology* 55 (6): 674–695.

page 38, Older elephant matriarchs that had lived through a previous drought thirty-five years earlier: Foley, C., N. Pettorelli, and L. Foley. 2008. "Severe Drought and Calf Survival in Elephants." *Biology Letters* 4: 541–544.

page 38, A mother chimpanzee teaches her offspring how to collect termites: Musgrave, S., D. Morgan, E. Lonsdorf, R. Mundry, and C. Sanz. 2016. "Tool Transfers Are a Form of Teaching among Chimpanzees." *Scientific Reports* 6 (1): 34783.

page 38, Inuit grandparents play an important role in teaching: Sarmiento, I. G. 2019. "Photos: How Families Eat in the Arctic: From an $18 Box of Cookies to Polar Bear Stew." NPR. https://www.npr.org/sections/goatsandsoda/2019/11/26/781679216/how-families-eat-in-the-arctic-from-an-18-box-of-cookies-to-polar-bear-stew.

page 39, Seven Grandfather Teachings: "Seven Grandfather Teachings." 2020. Nottawaseppi Huron Band of the Potawatomi. https://www.nhbpi.org/seven-grandfather-teachings/.

page 39, Taking part in group rituals can produce endorphins in our brains: Bundy, W. M. 2007. "Chapter 24: Religion II." In *Out of Chaos: Evolution from the Big Bang to Human Intellect.* Boca Raton, FL: Universal Publishers.

page 39, "Runner's high": Fuss, J., J. Steinle, L. Bindila, M. K. Auer, H. Kirchherr, B. Lutz, and P. Gass. 2015. "A runner's high depends on cannabinoid receptors in mice." *Proceedings of the National Academy of Sciences USA* 112 (42): 13105–13108.

page 39, Male rowers from Oxford University: Cohen, E. E., R. Ejsmond-Frey, N. Knight, and R. I. Dunbar. 2010. "Rowers' High: Behavioural Synchrony Is Correlated with Elevated Pain Thresholds." *Biology Letters* 6 (1): 106–108.

page 39, Group laughter can increase pain thresholds: Charles, S. J., V. van Mulukom, M. Farias, J. Brown, R. Delmonte, E. Maraldi, L. Turner, F. Watts, J. Watts, and R. Dunbar. 2020. "Religious Rituals Increase Social Bonding and Pain Threshold." PsyArXiv preprint.

page 39, Dysphoria: Whitehouse and Lanman, "The Ties That Bind Us," 2014.

page 39, Naghol: Neubauer, I. L. 2014. "Meet Vanuatu's Land-Diving Daredevils, Who Inspired Bungee Jumping." CNN. https://www.cnn.com/travel/article/vanuatu-land-divers/index.html.

page 40, In a study of fire walkers in a Mauritian Hindu community: Fischer, R., D. Xygalatas, P. Mitkidis, P. Reddish, P. Tok, I. Konvalinka, and J. Bulbulia. 2014. "The Fire-Walker's

notes

High: Affect and Physiological Responses in an Extreme Collective Ritual." see https://journals.plos.org/plosone/article?id=10.1371/journal.pone.0088355 *PLOS ONE* 9 (2): e88355.

page 41, Cooperation among captive wolves: Marshall-Pescini, S., J. F. L. Schwarz, I. Kostelnik, Z. Virányi, and F. Range. 2017. "Importance of a Species' Socioecology: Wolves Outperform Dogs in a Conspecific Cooperation Task." *Proceedings of the National Academy of Sciences USA* 114 (44): 11793–11798.

page 42, Group rituals often combine elements, like repeated vocalizations and synchronized, repeated body movements: Reddish, P., R. Fischer, and J. Bulbulia. 2013. "Let's Dance Together: Synchrony, Shared Intentionality and Cooperation." *PLOS ONE* 8 (8): e71182.

page 42, Studies have shown that group rituals involving synchronous movement: Fischer, R., R. Callander, P. Reddish, and J. Bulbulia. 2013. "How Do Rituals Affect Cooperation? An Experimental Field Study Comparing Nine Ritual Types." *Human Nature* 24 (2): 115–125.

COURTSHIP RITUALS

page 48, *The Rime of the Ancient Mariner:* Coleridge, S. T. 1900. *The Rime of the Ancient Mariner.* New York: Globe School Book Company.

page 49, The entire flock of flamingos engages in a group march: O'Connell-Rodwell, C. E., N. Rojek, T. C. Rodwell, and P. W. Shannon. 2004. "Artificially Induced Group Display and Nesting Behaviour in a Reintroduced Population of Caribbean Flamingo *Phoenicopterus ruber ruber.*" *Bird Conservation International* 14 (1): 55–62.

page 49, Flamingos secrete this pink substance from a preening gland near the tail: Krienitz, L. 2018. "The Lesser Flamingo." In *Lesser Flamingos.* Berlin, Heidelberg: Springer.

page 49, Female flamingos spend more time reapplying: Rose, P., and L. Soole. 2020. "What Influences Aggression and Foraging Activity in Social Birds? Measuring Individual, Group and Environmental Characteristics." *Ethology* 126 (9): 1–14.

page 50, The courtship ritual starts off with "head-flagging": O'Connell-Rodwell et al., "Artificially Induced Group Display and Nesting Behaviour in a Reintroduced Population of Caribbean Flamingo *Phoenicopterus ruber ruber,*" 2004.

page 50, Designed to show off the strength or health of a would-be mate: Fusani, L., J. Barske, L. D. Day, M. J. Fuxjager, and B. A. Schlinger. 2014. "Physiological Control of Elaborate Male Courtship: Female Choice for Neuromuscular Systems." *Neuroscience and Biobehavioral Reviews* 46 (4): 534–546.

page 51, In more promiscuous species: Hoglund, J., and R. V. Alatalo. 1995. *Leks. Monographs in Behavior and Ecology.* Princeton, NJ: Princeton University Press.

page 52, A male peacock's tail evolved: Darwin, C. 1871. *The Descent of Man: And Selection in Relation to Sex.* London: J. Murray.

page 52, Contrary to his theory of natural selection: Darwin, C. 1859. *On the Origin of Species by Means of Natural Selection, or, the Preservation of Favoured Races in the Struggle for Life.* London: J. Murray.

page 52, Case of the bower bird: Endler, J. A., J. Gaburro, and L. A. Kelley. 2014. "Visual Effects in Great Bowerbird Sexual Displays and Their Implications for Signal Design." *Proceedings of the Royal Society B: Biological Sciences* 281 (1783): 20140235.

page 52, The color red is a signal of high testosterone in males: Ligon, J. D., R. Thornhill, M. Zuk, and K. Johnson. 1990. "Male-Male Competition, Ornamentation and the Role of Testosterone in Sexual Selection in Red Jungle Fowl." *Animal Behaviour* 40 (2): 367–373.

page 52, They generate a scent and display a whole lot of swagger. Poole, J. H. 1987. "Rutting Behavior in African Elephants: The Phenomenon of Musth." *Behaviour* 102 (3–4): 283–316.

page 52, The red-capped manakin moon walk: Lindsay, W. R., J. T. Houck, C. E. Giuliano, and L. B. Day. 2015. "Acrobatic Courtship Display Coevolves with Brain Size in Manakins (*Pipridae*)." *Brain, Behavior and Evolution* 85 (1): 29–36.

page 53, Watching an ostrich mating dance: Bolwig, N. 1973. "Agonistic and Sexual Behavior of the African Ostrich (*Struthio camelus*)." *Condor* 75 (1): 100–105.

page 53, This group of birds evolved from a crow-like ancestor: Miles, M. C., and M. J. Fuxjager. 2018. "Synergistic Selection Regimens Drive the Evolution of Display Complexity in Birds of Paradise." *Journal of Animal Ecology* 87 (4): 1149–1159.

page 53, Male Wahnes's parotia, for example: Barske, J., B. A. Schlinger, M. Wikelski, and L. Fusani. 2011. "Female Choice for Male Motor Skills." *Proceedings of the Royal Society B: Biological Sciences* 278 (1724): 3523–3528.

page 54, Females choose males based on their motor skills: Schlinger, B. A., J. Barske, L. Day, L. Fusani, and M. J. Fuxjager. 2013. "Hormones and the Neuromuscular Control of Courtship in the Golden-Collared Manakin (*Manacus vitellinus*)." *Frontiers in Neuroendocrinology.* 34 (3): 143–156.

page 54, Study of the houbara bustard shows: Loyau, A., and F. Lacroix. 2010. "Watching Sexy Displays Improves Hatching Success and Offspring Growth through Maternal Allocation." *Proceedings of the Royal Society B: Biological Sciences* 277 (1699): 3453–3460.

page 54, The male booby presents the female with his lovely blue feet: Torres, R., and A. Velando. 2005. "Male Preference for Female Foot Colour in the Socially Monogamous Blue-Footed Booby, *Sula nebouxii*." *Animal Behaviour* 69 (1): 59–65.

page 55, The great bowerbird of northern Australia: Endler et al., "Visual Effects in Great Bowerbird Sexual Displays and Their Implications for Signal Design," 2014.

page 55, Sex-specific products can impact our hormone levels: Saad, G., and J. G. Vongas. 2009. "The Effect of Conspicuous Consumption on Men's Testosterone Levels." *Organizational Behavior and Human Decision Processes* 110 (2): 80–92.

page 55, Possession of a sports car influenced both men and women's perceptions: Saad, G., and T. Gill. 2014. "You Drive a Porsche: Women (Men) Think You Must Be Tall (Short), Intelligent and Ambitious." *NA: Advances in Consumer Research* 42: 808.

page 55, Women's behavior during their ovulation cycle: Zhuang, J., and J. Wang. 2014. "Women Ornament Themselves for Intrasexual Competition near Ovulation, but for Intersexual Attraction in Luteal Phase." *PLOS ONE* 9 (9): e106407.

page 56, Female elephants in estrus and musth rumbles: Poole, J. H., K. Payne, W. R. Langbauer Jr., and C. J. Moss. 1988. "The Social Contexts of Some Very Low Frequency Calls of African Elephants." *Behavioral Ecology and Sociobiology* 22: 385–392.

page 56, Young women perform a ritual dance with slices of apple: Wolchover, N. 2011. "Why Do We Kiss?" Live Science. https://www.livescience.com/33006-why-do-we-kiss.html.

page 56, Women unconsciously choose a mate based on a man's scent: Wedekind, C., and S. Füri. 1997. "Body Odour Preferences in Men and Women: Do They Aim for Specific MHC

notes

Combinations or Simply Heterozygosity?" *Proceedings of the Royal Society B: Biological Sciences* 264 (1387): 1471–1479.

page 57, Lions need to copulate to induce ovulation: Hunter, F. M., M. Petrie, M. Otronen, T. Birkhead, and A. P. Moller. 1993. "Why Do Females Copulate Repeatedly with One Male?" *Trends in Ecology and Evolution* 8 (1): 21–26.

page 58, Long-term relationships produce a host of benefits: Bales, K. L., W. A. Mason, C. Catana, S. R. Cherry, and S. P. Mendoza. 2007. "Neural Correlates of Pair-Bonding in a Monogamous Primate." *Brain Research* 1184: 245–253.

page 58, Sing-sing festival in Papua New Guinea: "Tumbuna Show: Small Sing Sing Festival, Big Cultural Experience." 2019. About Papau New Guinea. https://www.aboutpapuanewguinea.com/blog/tumbuna-show-small-sing-sing-festivals/.

page 59, This modern courtship ritual evolved from an ancient Roman festival: Bryner, J. 2013. "What's the Origin of Valentine's Day?" Live Science. https://www.livescience.com/32426-who-was-saint-valentine.html.

page 59, In the bundling ritual: Bailey, B. 2004. "From 'Bundling' to 'Hooking Up': Teaching the History of American Courtship." *OAH Magazine of History*, 3–4.

page 60, Courtship ritual called "visiting girls": Wolchover, "Why Do We Kiss?" 2011.

page 60, Couples say they don't need symbolic gestures: Braboy Jackson, P., S. Kleiner, C. Geist, and K. Cebulko. 2011. "Conventions of Courtship: Gender and Race Differences in the Significance of Dating Rituals." *Journal of Family Issues* 32 (5): 629–652.

GIFTING RITUALS

page 63, In 1977, he began sowing his seed in captivity: "Diego, the Galápagos Tortoise with a Species-Saving Sex Drive, Retires." 2020. BBC. https://www.bbc.com/news/world-latin-america-53062480.

page 63, *Nuptial gift:* Lewis, S. M., K. Vahed, J. M. Koene, L. Engqvist, L. F. Bussière, J. C. Perry, D. Gwynne, and G. U. C. Lehmann. 2014. "Emerging Issues in the Evolution of Animal Nuptial Gifts." *Biology Letters* 10 (7): 20140336.

page 64, If a female doesn't accept the nuptial gift of a male: Albo, M. J., and A. V. Peretti. 2015. "Worthless and Nutritive Nuptial Gifts: Mating Duration, Sperm Stored and Potential Female Decisions in Spiders." *PLOS ONE* 10 (6): e0129453.

page 65, A nuptial gift can also directly contribute to the female's sustenance by offsetting the energetic cost of nest-building, egg-laying, or having to nurse or feed offspring: ibid.

page 65, The gift can be symbolic: Sherry Jr., J. F. 1983. "Gift Giving in Anthropological Perspective." *Journal of Consumer Research* 10 (2): 157–168.

page 65, A gift is a form of communication: ibid.

page 65, A gift can offer protection, reciprocity, or even be a method of socialization: ibid.

page 66, Our earliest ancestors used gifts to demonstrate a man's ability to provide: van Schaik, C. P. 2016. "Chapter 22: Morality." In *The Primate Origins of Human Nature*, edited by Matt Cartmill and Kaye Brown, 351–362. Foundations of Human Biology. Hoboken, NJ: John Wiley & Sons.

page 66, Male chimps will often offer food gifts: Stevens, J. R., and I. C. Gilby. 2004. "A Conceptual Framework for Nonkin Food Sharing: Timing and Currency of Benefits." *Animal Behaviour* 67 (4): 603–614.

page 66, This typical Mayan gift makes sense: Lucero, L. J., and J. G. Cruz. 2020. "Reconceptualizing Urbanism: Insights from Maya Cosmology." *Frontiers in Sustainable Cities* 2 (1): 1–15.

page 66, Dowries are still expected in some cultures: Karasavvas, T. 2019. "Putting a Price on Marriage: The Long-standing Custom of Dowries." Ancient Origins. https://www.ancient-origins.net/history-ancient-traditions/putting-price-marriage-long-standing-custom-dowries-007222.

page 67, The act of gifting is a token of appreciation, of love and friendship, and a form of remediation: D'Costa, K. 2014. "The Obligation of Gifts." *Scientific American*. https://blogs.scientificamerican.com/anthropology-in-practice/the-obligation-of-gifts/.

page 67, The expectation of obligation: Sherry Jr., "Gift Giving in Anthropological Perspective," 1983.

page 67, There are two kinds of gifts: *transactional gifts* and *reciprocal gifts*: ibid.

page 67, A lioness might allow her adult daughter to steal: Stevens and Gilby, "A Conceptual Framework for Nonkin Food Sharing," 2004.

page 67, "There is no such thing as a free gift": Kidd, A. J. 1996. "Philanthropy and the 'Social History Paradigm.'" *Social History* 21 (2): 180–192.

page 68, The Hanging Gardens of Babylon: Cartwright, M. 2018. "Hanging Gardens of Babylon." Ancient History Encyclopedia. https://www.ancient.eu/Hanging_Gardens_of_Babylon/.

page 68, The Taj Mahal was built in Agra, India: "Taj Mahal." 1983. UNESCO. https://whc.unesco.org/en/list/252.

page 68, The first Fabergé egg: Morton, E. 2019. "The Mysterious Fate of the Romanov Family's Prized Easter Egg Collection." History. https://www.history.com/news/romanov-family-russia-mystery-faberge-easter-eggs.

page 68, Providing a neighbor nation with stone jars inscribed with a royal monogram: "A Brief History of Gift Giving and How to Do It Today." Occasion Station. https://www.occasionstation.com/a-brief-history-of-gift-giving-and-how-to-do-it-today/.

page 68, A modern example is the Statue of Liberty: "The French Connection." 2018. National Park Service. https://www.nps.gov/stli/learn/historyculture/the-french-connection.htm.

page 68, *Reciprocal gifts* refer to when related or unrelated individuals offer each other something: Stevens and Gilby. "A Conceptual Framework for Nonkin Food Sharing," 2004.

page 69, Food sharing has been demonstrated in such diverse species as nonhuman primates, elephants, lions, wolves, rats, birds and bats: ibid.

page 69, Captive African gray parrots were given a valuable token: Brucks, D., and A. M. P. von Bayern. 2020. "Parrots Voluntarily Help Each Other to Obtain Food Rewards." *Current Biology* 30 (2): 292–297.

page 69, A similar study conducted with three-year old children: Vaish, A., R. Hepach, and M. Tomasello. 2018. "The Specificity of Reciprocity: Young Children Reciprocate More Generously to Those Who Intentionally Benefit Them." *Journal of Experimental Child Psychology* 167: 336–353.

notes

page 69, One such behavior is "recruitment calling" to share food: Stevens and Gilby, "A Conceptual Framework for Nonkin Food Sharing," 2004.

page 70, Recruitment calling includes the "whoop" call among hyenas: Gersick, A. S., D. L. Cheney, J. M. Schneider, R. M. Seyfarth, and K. E. Holekamp. 2015. "Long-Distance Communication Facilitates Cooperation among Wild Spotted Hyaenas, *Crocuta crocuta*." *Animal Behaviour* 103: 107–116.

page 70, Cliff swallows commonly announce a food source: Brown, C. R. 1998. "Chapter 2: Whitetail." In *Swallow Summer*. Lincoln: University of Nebraska Press.

page 70, The house sparrow has a solution for the problem of attracting predators: Elgar, M. 1986. "House Sparrows Establish Foraging Flocks by Giving Chirrup Calls If the Resource Is Divisible." *Animal Behaviour* 34: 169–174.

page 70, Killer whales, for example, benefit by sharing a carcass of a seal: Guinet, C., L. G. Barrett-Lennard, and B. Loyer. 2000. "Co-ordinated Attack Behavior and Prey Sharing by Killer Whales at Crozet Archipelago: Strategies for Feeding on Negatively-Buoyant Prey." *Marine Mammal Science* 16 (4): 829–834.

page 70, The male praying mantis offers a food gift: Toft, S., and M. J. Albo. 2016. "The Shield Effect: Nuptial Gifts Protect Males against Pre-copulatory Sexual Cannibalism." *Biology Letters* 12: 20151082.

page 70, Chimpanzees share their food with those that harass them four times more often: Stevens, J. R. 2004. "The Selfish Nature of Generosity: Harassment and Food Sharing in Primates." *Proceedings of the Royal Society B: Biological Sciences* 271 (1538): 451–456.

page 70, Gifting down to a lower-ranking member of the group: Stevens and Gilby, "A Conceptual Framework for Nonkin Food Sharing," 2004.

page 71, The alpha male is more stressed than the beta male: Gesquiere, L. R., N. H. Learn, M. C. M. Simao, P. O. Onyango, S. C. Alberts, and J. Altmann. 2011. "Life at the Top: Rank and Stress in Wild Male Baboons." *Science* 333 (6040): 357–360.

page 71, Sharing a particularly abundant harvest or food cache encourages others to do the same: Standage, T. 2009. "Chapter 3: Food, Wealth, and Power." In *An Edible History of Humanity*. New York: Walker and Company.

page 71, This same dynamic exists among vampire bats: Wilkinson, G. S. 1984. "Reciprocal Food Sharing in the Vampire Bat." *Nature* 308: 181–184.

page 71, Social anthropologists have called the "gift-giving" paradox: Weaver, K., S. M. Garcia, and N. Schwarz. 2012. "The Presenter's Paradox: Figure 1." *Journal of Consumer Research* 39 (3): 445–460.

page 71, Chinese custom of giving either money or cigarettes: Huang, L. L., J. F. Thrasher, Y. Jiang, Q. Li, G. T. Fong, and A. C. Quah. 2012. "Incidence and Correlates of Receiving Cigarettes as Gifts and Selecting Preferred Brand Because It Was Gifted: Findings from the ITC China Survey." *BMC Public Health* 12: 996.

page 72, Considered a form of self-care, self-gifting is now common in Western societies: JamehBozorgi, M. J., and S. H. H. Dashtaki. 2014. "Motivations, Emotions, and Feelings of Self-Gifting Entrepreneurs: A Cross-Cultural Study." *Journal of Entrepreneurship, Business and Economics* 2 (2): 98–120.

page 72, In Eastern cultures, researchers believe that self-gifting may exist as a material means of obtaining an ideal self: Mick, D. G., and M. Demoss. 1990. "Self-Gifts: Phenomenological Insights from Four Contexts." *Journal of Consumer Research* 17 (3): 322–332.

page 72, The giver is asking forgiveness for some past transgression: Kelley, D. 1998. "The Communication of Forgiveness." *Communication Studies* 49 (3): 255–271.

page 72, If two chimpanzees have an altercation, they come together afterward and offer a hug: de Waal, F. B. M. 2006. "Bonobo Sex and Society." *SA Special Editions* 16 (3s): 14–21.

page 72, In bonobos, the gift of a sexual encounter reduces tensions surrounding food sharing: ibid.

page 73, Giving is more about the giver than the receiver: Sherry Jr., "Gift Giving in Anthropological Perspective," 1983.

page 73, Giving your partner a massage can be just as rewarding as receiving a massage: Naruse, S. M., P. L. Cornelissen, and M. Moss. 2018. "'To Give Is Better Than to Receive?' Couples Massage Significantly Benefits Both Partners' Wellbeing." *Journal of Health Psychology*: ISSN 1359-1053.

page 74, Belly rubs and ear scratches, seeing the pleasure the dog experiences, is rewarding: McGowan, R. T. S., C. Bolte, H. R. Barnett, G. Perez-Camargo, and F. Martin. 2018. "Can You Spare 15 Min? The Measurable Positive Impact of a 15-Min Petting Session on Shelter Dog Well-Being." *Applied Animal Behaviour Science* 203: 42–54.

page 74, The same is true of tickling a baby: Riem, M. M. E., M. H. van Ijzendoorn, M. Tops, M. A. S. Boksem, S. A. R. B. Rombouts, and M. J. Bakermans-Kranenburg. 2012. "No Laughing Matter: Intranasal Oxytocin Administration Changes Functional Brain Connectivity during Exposure to Infant Laughter." *Neuropsychopharmacology* 37 (5): 1257–1266.

page 74, Meerkats eat scorpions and have to learn how to eat them without getting stung: Thornton, A., and K. McAuliffe. 2006. "Teaching in Wild Meerkats." *Science* 313 (5784): 227–229.

page 74, A lioness will often present her cubs with wounded prey as a gift: Charlton, C. 2016. "Taught to Kill . . . by Mummy: Lion Cub Is Shown How to Hunt by Lioness after It Wounds Prey and Lets the Youngster Finish It Off." *Daily Mail*. https://www.dailymail.co.uk/news/article-3429745/Taught-kill-mummy-Lion-cub-shown-hunt-lioness-wounds-prey-lets-youngster-finish-off.html.

SPOKEN RITUALS

page 79, Battle cry is one of many ritualized vocalizations: Wiltermuth and Heath, "Synchrony and Cooperation," 2009.

page 80, The battle cry or call to arms serves the very important purpose: Legare and Watson-Jones, "The Evolution and Ontogeny of Ritual," 2015.

page 80, Vocal ritual with screams and hoots: Goodall, J. 2010. *Through a Window: My Thirty Years with the Chimpanzees of Gombe*. Boston: Houghton Mifflin Harcourt.

page 80, Engaging in these chants and aggressive vocalizations releases endorphins: Gelfand, M. J., N. Caluori, J. C. Jackson, and M. K. Taylor. 2020. "The Cultural Evolutionary Trade-off of Ritualistic Synchrony." *Philosophical Transactions of the Royal Society B: Biological Sciences* 375 (1805): 20190432.

notes

page 80, The roar of the crowd is a shared experience: Khairy, L. T., R. Barin, F. Demonière, C. Villemaire, M. Billo, J. Tardif, L. Macle, and P. Khairy. 2017. "Heart Rate Response in Spectators of the Montreal Canadiens Hockey Team." *Canadian Journal of Cardiology* 33 (12): 1633–1638.

page 80, Canadian doctors caution hockey fans to not take hockey games so seriously: ibid.

page 80, Experiencing similar physiological states as the athlete playing the game: Acharya, S., and S. Shukla. 2012. "Mirror Neurons: Enigma of the Metaphysical Modular Brain." *Journal of Natural Science, Biology and Medicine*. 3 (2): 118–124.

page 81, Elephants give "let's go" rumbles: O'Connell-Rodwell et al., "Antiphonal Vocal Bouts Associated with Departures in Free-Ranging African Elephant Family Groups," 2012.

page 81, Gorillas increasingly grunt: Harcourt, A. H., and K. J. Stewart. 1994. "Gorillas' Vocalizations during Rest Periods: Signals of Impending Departure?" *Behavior* 130 (1–2): 29–40.

page 81, Lion roars: Stander, P. E, and J. Stander. 1988. "Characteristics of Lion Roars in Etosha National Park." *Madoqua* 15 (4): 315–318.

page 81, Wolf howl: Harrington, F. H., and L. D. Mech. 1983. "Wolf Pack Spacing: Howling as a Territory-Independent Spacing Mechanism in a Territorial Population." *Behavioral Ecology and Sociobiology* 12 (2): 161–168.

page 81, Howler monkey calls at dawn and dusk: da Cunha et al., "The Roaring of Southern Brown Howler Monkeys (*Alouatta guariba clamitans*) as a Mechanism of Active Defence of Borders," 2007.

page 81, Elk bugle: Feighny, J. A., K. E. Williamson, and J. A. Clarke. 2006. "North American Elk Bugle Vocalizations: Male and Female Bugle Call Structure and Context." *Journal of Mammalogy* 87 (6): 1072–1077.

page 81, Assess the quality of a mate in red deer: Charlton, B. D., D. Reby, and K. McComb. 2007. "Female Red Deer Prefer the Roars of Larger Males." *Biology Letters* 3 (4): 382–385.

page 81, The red squirrel: Wilson, D. R., A. R. Goble, S. Boutin, M. M. Humphries, D. W. Coltman, J. C. Gorrell, J. Shonfield, and A. G. McAdam. 2015. "Red Squirrels Use Territorial Vocalizations for Kin Discrimination." *Animal Behaviour* 107: 79–85.

page 81, Soothing them with a lullaby: Cirelli, L. K., Z. B. Jurewicz, and S. E. Trehub. 2020. "Effects of Maternal Singing Style on Mother–Infant Arousal and Behavior." *Journal of Cognitive Neuroscience* 32 (7): 1213–1220.

page 81, Mammalian vocalizations are produced from rhythmic patterns of muscle contractions within the larynx: Titze, I. R. 2017. "Human Speech: A Restricted Use of the Mammalian Larynx." *Journal of Voice* 31 (2): 135–141.

page 81, Elephants might detect certain vocalizations as vibrations: O'Connell-Rodwell, C. 2007. "Keeping an 'Ear' to the Ground: Seismic Communication in Elephants." *Physiology* 22: 287–294.

page 81, Vibrating material like a leaf: Casas, J., C. Magal, and J. Sueur. 2007. "Dispersive and Non-dispersive Waves through Plants: Implications for Arthropod Vibratory Communication." *Proceedings of the Royal Society B: Biological Sciences* 274 (1613): 1087–1092.

page 81, Human and nonhuman animals can detect vibrations through bone-conduction: Puria, S., and J. J. Rosowski. 2012. "Bekesy's Contributions to Our Present Understanding of Sound Conduction to the Inner Ear." *Hearing Research* 293 (1–2): 21–30.

page 82, The evolution of human language might be due in part to the FoxP2 gene: Mozzi, A., D. Forni, M. Clerici, U. Pozzoli, S. Mascheretti, F. R. Guerini, S. Riva, N. Bresolin, R. Cagliani, and M. Sironi. 2016. "The Evolutionary History of Genes Involved in Spoken and Written Language: Beyond FOXP2." *Scientific Reports* 6 (1): 22157.

page 82, For humans, spoken language has long been a necessary element of social cooperation: van Schaik, C. P. 2016. "Chapter 22: Morality." In *The Primate Origins of Human Nature*, edited by Matt Cartmill and Kaye Brown, 351–362. Foundations of Human Biology. Hoboken, NJ: John Wiley & Sons.

page 82, Scientists attribute the evolution of language to the evolution of more sophisticated tools: Szamado, S., and E. Szathmary. 2006. "Selective Scenarios for the Emergence of Natural Language." *Trends in Ecology and Evolution* 21 (10): 555–561.

page 82, Social animals are impacted physiologically when they create and hear vocalizations: Seltzer, L. J., T. E. Ziegler, and S. D. Pollak. 2010. "Social Vocalizations Can Release Oxytocin in Humans." *Proceedings of the Royal Society B: Biological Sciences* 277 (1694): 2661–2666.

page 83, If the sender has positive intentions—say, a man serenading a lover in courtship—the hormone oxytocin gets released in both the singer and the listener: ibid.

page 83, Like the swamp boubou, males and females sing in a remarkably coordinated duet ritual: Thorpe, W. H. 1973. "Duet-Singing Birds." *Scientific American* 229 (2): 70–79.

page 83, It causes her ovarian follicles to double in size: Lehrman, D. S., and M. Friedman. 1969. "Auditory Stimulation of Ovarian Activity in the Ring Dove (*Streptopelia risoria*)." *Animal Behaviour* 17 (3): 494–497.

page 83, Visual displays stimulating a higher birth rate in houbara bustards: Loyau and Lacroix. "Watching Sexy Displays Improves Hatching Success and Offspring Growth through Maternal Allocation," 2010.

page 83, When domestic pigs are separated from their companions: Shrader, L., and D. Todt. 1998. "Vocal Quality Is Correlated with Levels of Stress Hormones in Domestic Pigs." *Ethology* 104 (10): 859–876.

page 84, Low-frequency sounds often correlate with larger body size, or perceived body size: Bowling, D. L., M. Garcia, J. C. Dunn, R. Ruprecht, A. Stewart, K. H. Frommolt, and W. T. Fitch. 2017. "Body Size and Vocalization in Primates and Carnivores." *Scientific Reports* 7: 41070.

page 84, By projecting a lower-frequency call, the signaler provides information about their size: ibid.

page 84, Or a low-pitched male call can serve to attract a mate: ibid.

page 84, A wolf emitting a bark-howl is sending the message: McCarley, H. 1978. "Vocalizations of Red Wolves (*Canis rufus*)." *Journal of Mammalogy* 59 (1): 27–35.

page 84, A male lion entering an area will roar: McComb, K., A. Pusey, C. Packer, and J. Grinnell. 1993. "Female Lions Can Identify Potentially Infanticidal Males from Their Roars." *Proceedings of the Royal Society B: Biological Sciences* 252: 59–64.

page 84, On the other hand, some frogs cheat: Tan, W. H., C. G. Tsai, C. Lin, and Y. K. Lin. 2014. "Urban Canyon Effect: Storm Drains Enhance Call Characteristics of the Mientien Tree Frog." *Journal of Zoology* 294 (2): 77–84.

page 84, A speaker at a press conference, professional presentation, or board meeting may use a lower voice: Nikitina, A. 2011. *Successful Public Speaking*. bookboon.com.

notes

page 85, A female red deer can decide on her mate by assessing the volume of the male roars: Charlton, et al., "Female Red Deer Prefer the Roars of Larger Males," 2007.

page 85, The wahoo calls of the male baboon: Bergman, T. J., J. C. Beehner, D. L. Cheney, R. M. Seyfarth, and P. L. Whitten. 2005. "Interactions in Male Baboons: The Importance of Both Males' Testosterone." *Behavioral Ecology and Sociobiology* 59 (4): 480–489.

page 85, In the male common loon, a lengthening of his yodel vocalization: Walcott, C., D. Evers, M. Froehler, and A. Krakauer. 1999. "Individuality in "Yodel" Calls Recorded from a Banded Population of Common Loons, *Gavia immer*." *Bioacoustics* 10 (2–3): 101–114.

page 85, For the red squirrel, territorial vocalizations: Wilson et al., "Red Squirrels Use Territorial Vocalizations for Kin Discrimination," 2015.

page 85, The territorial "great call" of the female, white-handed gibbon: Terleph, T. A., S. Malaivijitnond, and U. H. Reichard. 2016. "Age Related Decline in Female Lar Gibbon Great Call Performance Suggests That Call Features Correlate with Physical Condition." *BMC Evolutionary Biology* 16: 4.

page 85, Vocal rituals make it possible for mothers and offspring to identify and find each other in a crowd: Thierry, A., P. Jouventin, and I. Charrier. 2015. "Mother Vocal Recognition in Antarctic Fur Seal *Arctocephalus gazella* Pups: A Two-Step Process." *PLOS ONE* 10 (9): e0134513.

page 85, A human baby learns to recognize Mom's voice from inside the womb: Kisilevsky, B. S., S. M. J. Hains, K. Lee, X. Xie, H. Huang, H. H. Ye, K. Zhang, and Z. Wang. 2003. "Effects of Experience on Fetal Voice Recognition." *Psychological Science* 14 (3): 220–224.

page 85, Coordinated vocal rituals evolved for predators to keep track of one another during a hunt: Petak, I. 2010. "Patterns of Carnivores' Communication and Potential Significance for Domestic Dogs." *Periodicum Biologorum* 112 (2): 127–132.

page 86, Whales communicate very long distances by taking advantage of the SOFAR (sound fixing and ranging) channel deep in the ocean: "What Is SOFAR?" 2018. NOAA. https://oceanservice.noaa.gov/facts/sofar.html.

page 86, This is why so many animals communicate at dawn and dusk: Brown, T. J., and P. Handford. 2002. "Why Birds Sing at Dawn: The Role of Consistent Song Transmission." *Ibis* 145 (1): 120–129.

page 86, Vibrations from an elephant's low-frequency rumble may be heard much farther through the ground than through the air: O'Connell-Rodwell, C. E., B. T. Arnason, and L. A. Hart. 2000. "Seismic Properties of Asian Elephant (*Elephas maximus*) Vocalizations and Locomotion." *Journal of the Acoustical Society of America* 108 (6): 3066–3072.

page 86, The first known long-distance instrument was the bullroarer: "Australian Bullroarer." 2018. Wake Forest University. https://moa.wfu.edu/2018/07/australian-bullroarer/.

page 87, Native Americans used song and drumming rituals, along with smoke signals: Bryant, C. W. 2008. "How Do You Send a Smoke Signal?" Howstuffworks. https://adventure.howstuffworks.com/survival/wilderness/how-to-send-smoke-signal.htm.

page 87, Traditional talking drums of West Africa were used for a similar purpose: Carrington, J. F. 1949. *Talking Drums of Africa*. London: Carey Kingsgate Press.

page 87, Aboriginals in Australia invented the didgeridoo, which generates low-frequency sounds: Koumoulas, M. 2018. "Didgeridoo Notation." Master of Arts, Music, York University.

page 87, In the first account of yodeling, in 397 CE: Platenga, B. 1999. "Will There Be Yodeling in Heaven?" *American Music Research Center Journal* 9: 107–138.

page 88, West African Pygmies have a yodel ritual: Pemunta, N. V. 2018. "Fortress Conservation, Wildlife Legislation and the Baka Pygmies of Southeast Cameroon." *GeoJournal* 84 (4): 1035–1055.

page 88, Almost every culture and musical genre—from classical, rock, R&B, jazz, and country to opera—uses yodeling: Platenga, "Will There Be Yodeling in Heaven?," 1999.

page 88, Even Adele uses a form of yodeling in her songs: Berger, M. "Adele's 25—Track by Track." Harvard Institute of Politics. https://iop.harvard.edu/get-involved/harvard-political-review/adele%E2%80%99s-25-track-track.

page 88, The practice was banned in the presence of Swiss mercenary soldiers: Rechsteiner, A. 2019. "Homesick for the Mountains." SWI swissinfo.ch. https://www.swissinfo.ch/eng/swiss-national-museum_homesick-for-the-mountains/45282814.

page 89, Our bodies benefit from vocal exchanges: Seltzer, L. J., A. R. Prososki, T. E. Ziegler, and S. D. Pollak. 2012. "Instant Messages vs. Speech: Hormones and Why We Still Need to Hear Each Other." *Evolution and Human Behavior* 33 (1): 42–45.

page 89, Studies show that spoken rituals increase our level of commitment: Hobson et al., "The Psychology of Rituals," 2018.

page 89, The simple act of sharing our thoughts to another person or a group: Huron, D. 2001. "Is Music an Evolutionary Adaptation?" *Annals of the New York Academy of Sciences* 930: 43–61.

page 89, Singing out loud: Moss, H., J. Lynch, and J. O'Donoghue. 2018. "Exploring the Perceived Health Benefits of Singing in a Choir: An International Cross-Sectional Mixed-Methods Study." *Perspectives in Public Health* 138 (3): 160–168.

page 89, Singing in a group lowers cortisol levels and increases the expression of oxytocin: van Schaik, C. P. 2016. *The Primate Origins of Human Nature*, edited by Matt Cartmill and Kaye Brown. Foundations of Human Biology. Hoboken, NJ: John Wiley & Sons.

page 90, Singing also reduces depression and feelings of loneliness: Raglio, A., L. Attardo, G. Gontero, S. Rollino, E. Groppo, and E. Granieri. 2015. "Effects of Music and Music Therapy on Mood in Neurological Patients." *World Journal of Psychiatry* 5 (1): 68–78.

page 90, The elation we feel when singing—in the car, in a choir, in a karaoke bar with strangers: Horn, S. 2013. "Singing Changes Your Brain." *Time*. https://ideas.time.com/2013/08/16/singing-changes-your-brain/.

page 90, Singing is now being used to manage the psychological side effects of a range of disorders: Jasemi, M., S. Aazami, and R. E. Zabihi. 2016. "The Effects of Music Therapy on Anxiety and Depression of Cancer Patients." *Indian Journal of Palliative Care* 22 (4): 455–458.

page 90, Music therapy can reduce anxiety and improve quality of life: Puhan, M. A., A. Suarez, C. Lo Cascio, A. Zahn, M. Heitz, and O. Braendli. 2006. "Didgeridoo Playing as Alternative Treatment for Obstructive Sleep Apnoea Syndrome: Randomised Controlled Trial." *BMJ* 332 (7536): 266–270.

page 90, Many scientists believe music and song are not evolutionary adaptive behaviors: Huron, "Is Music an Evolutionary Adaptation?," 2001.

page 90, He suggests that pleasure-seeking behaviors: ibid.

page 90, The earliest known musical instrument, found in Slovenia: ibid.

page 90, Huron reviews eight theories for how music may have evolved: ibid.

page 91, Music has been used to pass along information: Somerville, M., L. Tobin, and J. Tobin. 2019. "Walking Contemporary Indigenous Songlines as Public Pedagogies of Country." *Journal of Public Pedagogies* 4: 13–27.

page 91, Music making may in fact be in our genes: Callaway, E. 2007. "Music Is in Our Genes." *Nature.*

page 91, Not only singing but listening to music is a pleasurable activity: Ferreri, L., E. Mas-Herrero, R. J. Zatorre, P. Ripolles, A. Gomez-Andres, H. Alicart, G. Olive, J. Marco-Pallares, R. M. Antonijoan, M. Valle, J. Riba, and A. Rodriguez-Fornells. 2019. "Dopamine Modulates the Reward Experiences Elicited by Music." *Proceedings of the National Academy of Sciences USA* 116 (9): 3793–3798.

page 91, Music arouses strong emotions: Huron, "Is Music an Evolutionary Adaptation?" 2001.

page 91, Alzheimer's disease experiences from an increase in melatonin levels: Kumar, A. M., F. Tims, D. G. Cruess, M. J. Mintzer, G. Ironson, D. Loewenstein, R. Cattan, J. B. Fernandez, C. Eisdorfer, and M. Kumar. 1999. "Music Therapy Increases Serum Melatonin Levels in Patients with Alzheimer's Disease." *Alternative Therapies in Health and Medicine* 5 (6): 49–57.

page 91, All religions use music, singing, and chanting to generate a sense of awe and humility: van Schaik, *The Primate Origins of Human Nature*, 2016.

page 91, Chanting the Sanskrit word *Om* during meditation: Dudeja, J. P. 2017. "Scientific Analysis of Mantra-Based Meditation and Its Beneficial Effects: An Overview." *International Journal of Advanced Scientific Technologies in Engineering and Management Sciences* 3 (6): 21–26.

UNSPOKEN RITUALS

page 98, Six-year research project to understand wolf society: Dutcher, J., and J. Dutcher. 2020. Living with Wolves. https://www.livingwithwolves.org/about-wolves/social-wolf/.

page 98, Dominance can fluctuate in wolf packs, depending on the very fluid social dynamics: ibid.

page 99, Unspoken rituals of dominance provide structure to societies: Hermann, H. R. 2017. "Chapter 1: Defining Dominance and Aggression." In *Dominance and Aggression in Humans and Other Animals: The Great Game of Life.* Amsterdam: Elsevier.

page 99, Elephant matriarch, members of the group entrust their safety to her: Moss, C. J. 2001. "The Demography of an African Elephant (*Loxodonta africana*) Population in Amboseli, Kenya." *Journal of Zoology* 255 (2): 145–156.

page 99, The olive baboon, however, leadership is more democratic: Strandburg-Peshkin, A., D. R. Farine, I. D. Couzin, and M. C. Crofoot. 2015. "Shared Decision-Making Drives Collective Movement in Wild Baboons." *Science* 348 (6241): 1358–1361.

page 99, Certain postures fill us with confidence: Carney, D. R., A. J. Cuddy, and A. J. Yap. 2010. "Power Posing: Brief Nonverbal Displays Affect Neuroendocrine Levels and Risk Tolerance." *Psychological Science* 21 (10): 1363–1368.

page 100, In the *Anolis* lizard, the males are constantly strutting about: Johnson, M. A., B. K. Kircher, and D. J. Castro. 2018. "The Evolution of Androgen Receptor Expression and Behavior in Anolis Lizard Forelimb Muscles." *Journal of Comparative Physiology A: Neuroethology, Sensory, Neural, and Behavioral Physiology* 204 (1): 71–79.

page 100, It turns out that the pose that we strike while seated in an important business meeting: Cuddy, A. J. C., C. A. Wilmuth, A. J. Yap, and D. R. Carney. 2015. "Preparatory Power Posing Affects Nonverbal Presence and Job Interview Performance." *Journal of Applied Psychology* 100 (4): 1286–1295.

page 100, Spike of testosterone accompanies power poses: Carney et al., "Power Posing," 2010.

page 100, Both of these behaviors correspond with higher testosterone levels in the signaler: ibid.

page 100, Elephants: O'Connell-Rodwell, C. 2017. "Elephant Country Blog 4: The Unseated Ozzie." *National Geographic Society Newsroom* (blog), *National Geographic*. https://blog.nationalgeographic.org/2017/07/25/elephant-country-blog-4-the-unseated-ozzie/.

page 100, Orangutans exhibit these behaviors, there is a decrease in testosterone expression: Emery Thompson, M., A. Zhou, and C. D. Knott. 2012. "Low Testosterone Correlates with Delayed Development in Male Orangutans." *PLOS ONE* 7 (10): e47282.

page 100, Holding one's head down and making minimal eye contact can convey insecurity: Pease, A., and B. Pease. 2004. *The Definitive Book of Body Language*. New York: Bantam Books.

page 101, Body language and facial expressions—all impact our physical and mental well-being: Segal, J., M. Smith, L. Robinson, and G. Boose. 2019. "Nonverbal Communication." HelpGuide. https://www.helpguide.org/articles/relationships-communication/nonverbal-communication.htm.

page 101, Gestures may even impact our reproductive health: House, L. D., et al. "Competence as a Predictor of Sexual and Reproductive Health Outcomes for Youth: A Systematic Review." *Journal of Adolescent Health* 46 (3): S7–S22.

page 101, For a male elephant in musth, his far-reaching olfactory cues serve three purposes: Poole, J. H. 1987. "Rutting Behavior in African Elephants: The Phenomenon of Musth." *Behaviour* 102 (3/4): 283–316.

page 104, Character is an important determining feature of dominance: Sarros, J. C., B. Cooper, and J. C. Santora. 2007. "The Character of Leadership." *Ivey Business Journal* 71 (5): 1–9.

page 104, Greg had a fascinating balance between carrot and stick: O'Connell, *Elephant Don*, 2012.

page 105, When young males were introduced to the parks: Slotow, R., G. van Dyk, J. Poole, B. Page, and A. Klocke. 2000. "Older Bull Elephants Control Young Males." *Nature* 408 (6811): 425–426.

page 105, The role of aggression and hormones in male dominance hierarchies in wild elephants: O'Connell-Rodwell, C. E., J. D. Wood, C. Kinzley, T. C. Rodwell, C. Alarcon, S. K. Wasser, and R. Sapolsky. 2011. "Male African Elephants (*Loxodonta africana*) Queue When the Stakes Are High." *Ethology Ecology and Evolution* 23 (4): 388–397.

page 105, Women living in a college dorm synchronize: Weller, L., A. Weller, H. Koresh-Kamin, and R. Ben-Shoshan. 1999. "Menstrual Synchrony in a Sample of Working Women." *Psychoneuroendocrinology* 24 (4): 449–459.

page 105, Synchronized cycles, and thus synchronized births, are very common in social animals: Ims, R. A. 1990. "The Ecology and Evolution of Reproductive Synchrony." *Trends in Ecology and Evolution* 5: 135–140.

page 105, Penguins: Ancel, A., M. Beaulieu, and C. Gilbert. 2013. "The Different Breeding Strategies of Penguins: A Review." *Comptes Rendus Biologies* 336 (1): 1–12.

notes

page 105, Flamingos: Martinez, F., and B. Durham. "Advantages of Reproductive Synchronization in the Caribbean Flamingo." Final Paper, Stanford University. https://socobilldurham.sites.stanford.edu/sites/g/files/sbiybj10241/f/soco_-_advantages_of_reproductive_synchronization_in_the_caribbean_flamingo.pdf

page 105, Synchronized births in antelope: Sekulic, R. 1978. "Seasonality of Reproduction in the Sable Antelope." *African Journal of Ecology* 16 (3): 177–182.

page 106, Predators such as lions benefit from synchronous births: Bertram, B. C. R. 2009. "Social Factors Influencing Reproduction in Wild Lions." *Journal of Zoology* 177 (4): 463–482.

page 106, We know that children need role models: Ahrens, K. R., D. L. Dubois, M. Garrison, R. Spencer, L. P. Richardson, and P. Lozano. 2011. "Qualitative Exploration of Relationships with Important Non-parental Adults in the Lives of Youth in Foster Care." *Children and Youth Services Review* 33 (6): 1012–1023.

page 106, Rituals that provide a bridge from adolescence to adulthood: Davis, J. "Wilderness Rites of Passage: Healing, Growth, and Initiation." School of Lost Borders. http://www.schooloflostborders.org/content/wilderness-rites-passage-healing-growth-and-initiation-john-davis-phd.

page 106, The bar and bat mitzvah in Jewish culture are examples of such rites of passage: ibid.

page 106, Chimpanzees do, indeed, smile while laughing: Davila-Ross, M., G. Jesus, J. Osborne, and K. A. Bard. 2015. "Chimpanzees (*Pan troglodytes*) Produce the Same Types of 'Laugh Faces' When They Emit Laughter and When They Are Silent." *PLOS ONE* 10 (6): e0127337.

page 107, A smile is good for your health—so is laughter: Dimberg, U., and S. Soderkvist. 2010. "The Voluntary Facial Action Technique: A Method to Test the Facial Feedback Hypothesis." *Journal of Nonverbal Behavior* 35: 17–33.

page 107, Parent and child get all warm and fuzzy when they look deeply into each other's eyes: Feldman, R., A. Weller, O. Zagoory-Sharon, and A. Levine. 2007. "Evidence for a Neuroendocrinological Foundation of Human Affiliation." *Psychological Science* 18 (11): 965–970.

page 107, The same thing happens between loved ones or even between a dog and its owner: Handlin, L., E. Hydbring-Sandberg, A. Nilsson, M. Ejdebäck, A. Jansson, and K. Uvnäs-Moberg. 2015. "Short-Term Interaction between Dogs and Their Owners: Effects on Oxytocin, Cortisol, Insulin and Heart Rate—An Exploratory Study." *Anthrozoös* 24 (3): 301–315.

page 107, Gaze serves as an important "tell": Bania, A. E., and E. E. Stromberg. 2013. "The Effect of Body Orientation on Judgments of Human Visual Attention in Western Lowland Gorillas (*Gorilla gorilla gorilla*)." *Journal of Comparative Psychology*, 127(1): 82–90.

page 107, Theory of mind has also been demonstrated in chimpanzees, bonobos and orangutans: Kano, F., Krupenye, C., Hirata, S., Tomonaga, M., and J. Call. 2019. "Great Apes Use Self-Experience to Anticipate an Agent's Action in a False-Belief Test." *Proceedings of the National Academy of Sciences* 116 (42) 20904–20909.

page 107, A powerful form of communication called American Sign Language: "American Sign Language." 2019. NIH. https://www.nidcd.nih.gov/health/american-sign-language.

page 108, The first orangutan to learn sign language: Omarzu, T. 2014. "New Documentary Tells Story of Orangutan Who Learned Sign Language at UTC." *Chattanooga Times Free Press*. https://www.timesfreepress.com/news/local/story/2014/jul/23/program-tells-chanteks-story/262460/.

page 109, A gorilla named Michael, who had been born in the wild: "Michael's Story." Gorilla Foundation. https://www.koko.org/conservation/michaels-story/.

page 110, Watching Michael tell his horrific tale through sign on Kokoflix is very humbling: ibid.

page 111, This knowledge of specific migratory routes and resources gets passed from the matriarch to all the adult females in the family: Foley, C., Pettorelli, N., and L. Foley. 2008. "Severe Drought and Calf Survival in Elephants." *Biological Letters* 4: 541–544.

page 111, Donna could imagine an item in her mind: O'Connell, "The Emotional Elephant," in *Elephant Don*, 2012.

page 112, Border collie named Chaser learned the words for over a thousand objects: Lee, A. 2013. "Smart Dog: Border Collie Learns Language, Grammar." *USA Today*. https://www.usatoday.com/story/news/nation/2013/11/24/smart-dog-border-collie-learns-language-grammar/3691967/.

page 112, A deaf dog named Blue: "Deaf Shelter Dog Masters Doggie Sign Language to Impress Her Future Family." *People*. https://people.com/pets/deaf-shelter-dog-learns-sign-language/.

page 112, Honest, subconscious signaling: Pentland, S. 2008. *Honest Signals: How They Shape Our World*. Cambridge, MA: MIT Press.

page 113, Pentland applied this formula to speed dating: Madan, A., R. Caneel, and S. Pentland. 2004. "Voices of Attraction." MIT Media Laboratory Technical Note No. 584.

page 114, The vibration of dancing can have a profound impact: Rooke, J. 2014. "The Restorative Effects of Ecstatic Dance: A Qualitative Study." BA (Hons) in Social Science, Dublin Business School.

page 114, This dance ritual has been practiced throughout human history: Berggren, K. 1998. *Circle of Shaman: Healing through Ecstasy, Rhythm, and Myth*. Rochester, VT: Destiny Books.

page 114, Ecstatic dance was incorporated into modern dance in the 1970s by Gabrielle Roth and has had a following in today's club cultures in the form of raves: ibid.

page 114, Tactile gestures such as grooming are another set of unspoken rituals: Crockford, C., R. M. Wittig, K. Langergraber, T. E. Ziegler, K. Zuberbuhler, and T. Deschner. 2013. "Urinary Oxytocin and Social Bonding in Related and Unrelated Wild Chimpanzees." *Proceedings of the Royal Society B: Biological Sciences* 280 (1755): 20122765.

page 114, These actions release the bonding hormone oxytocin, which promotes feelings of trust: Tierney, R. 2016. "The Power of Touch." *Telegraph*. https://www.telegraph.co.uk/beauty/skin/youthful-vitality/the-power-of-touch/.

page 114, Skin-to-skin contact during cuddling signals the adrenal gland to stop producing the stress hormone cortisol, which boosts our immune response and makes us healthier: ibid.

page 114, Touch releases serotonin and dopamine, which improve mood and curb depression: ibid.

page 114, Promoting better sleep: Grewen, K. M., B. J. Anderson, S. S. Girdler, and K. C. Light. 2003. "Warm Partner Contact Is Related to Lower Cardiovascular Reactivity." *Behavioral Medicine* 29 (3): 123–130.

page 115, Couples who cuddle more have stronger, healthier relationships: van Anders, S. M., R. S. Edelstein, R. M. Wade, and C. R. Samples-Steele. 2013. "Descriptive Experiences and Sexual vs. Nurturant Aspects of Cuddling between Adult Romantic Partners." *Archives of Sexual Behavior* 42: 553–560.

page 115, Cuddling is so critical for newborns: Walsh, K. 2019. "Calling All 'Cuddlers' to Volunteer." Denver CBS Local. https://denver.cbslocal.com/2019/01/15/cuddlers-volunteer-nicu-uchealth-university-colorado-hospital/.

notes

PLAY RITUALS

page 120, Play is actually very important to physical and social development and even to survival: Spinka, M., R. C. Newberry, and M. Beckoff. 2001. "Mammalian Play: Training for the Unexpected." *Quarterly Review of Biology* 76 (2): 141–168.

page 120, When mothers play with their young, both male and female foals develop better conditioning: Cameron, E. Z., W. L. Linklater, K. J. Stafford, and E. O. Minot. 2008. "Maternal Investment Results in Better Foal Condition through Increased Play Behaviour in Horses." *Animal Behaviour* 76 (5): 1511–1518.

page 121, Play behaviors are, by nature, ritualized or exaggerated forms of routine behaviors: van Schaik, C. P. 2016. "Chapter 16: Growth and Development." In *The Primate Origins of Human Nature*, edited by Matt Cartmill and Kaye Brown, 251–262. Foundations of Human Biology. Hoboken, NJ: John Wiley & Sons.

page 121, The play environment offers a special zone of protection: Spinka et al., "Mammalian Play," 2001.

page 121, Most humans and nonhuman animals participate in *social*, *locomotive*, and *object* play: Burghardt, G. M. 2005. *The Genesis of Animal Play: Testing the Limits*. Cambridge, MA: MIT Press.

page 121, Play includes all three, such as pretend role-playing with others while incorporating objects: van Schaik, "Chapter 16: Growth and Development," 2016.

page 122, This posture, known as the play bow, is an encoded invitation: Bekoff, M. 1995. "Play Signals as Punctuation: The Structure of Social Play in Canids." *Behaviour* 132 (5/6): 419–429.

page 122, In wolves, the play bow is often initiated by the omega: Dutcher, J. & J. Dutcher. 2013. *The Hidden Life of Wolves*. Washington, DC: National Geographic.

page 122, The goal isn't necessarily to "win," but to practice and improve essential skills: Spinka et al., "Mammalian Play," 2001.

page 122, Locomotive play—running, walking, jumping, and pouncing—develops life-long motor skills: ibid.

page 122, Giraffes play fight: Leuthold, B. M., and W. Leuthold. 1978. "Daytime Activity Patterns of Gerenuk and Giraffe in Tsavo National Park, Kenya." *African Journal of Ecology* 16 (4): 231–243.

page 123, Male rats also exhibit a form of play that is both practice for competitive fighting: Oliveira, A. F. S., A. O. Rossi, L. F. R. Silva, M. C. Lau, and R. E. Barreto. 2009. "Play Behaviour in Nonhuman Animals and the Animal Welfare Issue." *Journal of Ethology* 28 (1): 1–5.

page 123, Play is a low-cost, low-risk way to learn new behaviors: Pellegrini, A. D., D. Dupuis, and P. K. Smith. 2007. "Play in Evolution and Development." *Developmental Review* 27 (2): 261–276.

page 123, All forms of play peak during the juvenile stage: Oliveira et al., "Play Behaviour in Nonhuman Animals and the Animal Welfare Issue," 2009.

page 123, Animals are stressed or don't have enough to eat, or if the environment isn't safe: ibid.

page 123, It ignites innovation and exploration, and it promotes risk-taking and a flexible mindset: van Schaik, "Chapter 16: Growth and Development," 2016.

page 124, Cautionary tale of *Homo erectus*: Solly, M. 2018. "Laziness May Have Contributed to the Decline of *Homo erectus*." *Smithsonian Magazine*. https://www.smithsonianmag.com/smart-news/laziness-may-have-contributed-downfall-homo-erectus-180969983.

page 124, Risk-taking behaviors between a traditional desert society, the Himba of Namibia: Pope, S. M., J. Fagot, A. Meguerditchian, D. A. Washburn, and W. D. Hopkins. 2019. "Enhanced Cognitive Flexibility in the Semi-Nomadic Himba." *Journal of Cross-Cultural Psychology* 50 (1): 47–62.

page 125, Researchers gave one group of mice, both young and old, a running wheel in their habitat: van Praag, H., T. Shubert, C. Zhao, and F. H. Gage. 2005. "Exercise Enhances Learning and Hippocampal Neurogenesis in Aged Mice." *Journal of Neuroscience* 25 (38): 8680–8685.

page 125, Another study compared rat pups that were allowed to play to a group that wasn't: Einon, D. F., M. J. Morgan, and C. C. Kibbler. 1987. "Brief Periods of Socialization and Later Behavior in the Rat." *Developmental Psychobiology* 11 (3): 213–225. PMID: 658602.

page 125, Direct relationship between playfulness and cognitive development: Trevlas, E., O. Matsouka, and E. Zachopoulou. 2010. "Relationship between Playfulness and Motor Creativity in Preschool Children." *Early Child Development and Care* 173 (5): 535–543.

page 126, Parental involvement in play can also help develop a deep bond: Pellegrini et al., "Play in Evolution and Development," 2007.

page 126, Severe play deprivation in childhood leads to abnormal neurological development: Panksepp, J. 2007. "Can Play Diminish ADHD and Facilitate the Construction of the Social Brain?" *Journal of the Canadian Academy of Child and Adolescent Psychiatry* 16 (2): 57–66. PMID: 18392153.

page 126, "In Defense of Play": Gopnik, A. 2016. "In Defense of Play." *Atlantic.* https://www.theatlantic.com/education/archive/2016/08/in-defense-of-play/495545/.

page 126, When kids play video games together, this can strengthen bonds: Hickerson, B., and A. J. Mowen. 2012. "Behavioral and Psychological Involvement of Online Video Gamers: Building Blocks or Building Walls to Socialization?" *Society and Leisure* 35 (1): 79–103.

page 126, Video games can be highly addictive and lead to isolation and reduced socialization: Zamani, E., A. Kheradmand, M. Cheshmi, A. Abedi, and N. Hedayati. 2010. "Comparing the Social Skills of Students Addicted to Computer Games with Normal Students." *Journal of Addiction and Health* 2 (3–4): 59–65.

page 127, Being silly is actually a highly adaptive behavior: Gopnik, "In Defense of Play," 2016.

page 127, Play is also a major source of stress release: Robinson, L., M. Smith, J. Segal, and J. Shubin. 2019. "The Benefits of Play for Adults." HelpGuide. https://www.helpguide.org/articles/mental-health/benefits-of-play-for-adults.htm.

page 127, Many companies and corporations strive to create a sense of community and innovation: Mack, S. "Are Company Retreats Good for Productivity?" *Houston Chronicle.* https://smallbusiness.chron.com/company-retreats-good-productivity-37136.html.

page 128, Organized sports qualify as play and offer many similar benefits: Johnson, R. L., and P. Stanford. 2002. *Strength for Their Journey: 5 Essential Disciplines African-American Parents Must Teach Their Children and Teens.* New York: Harlem Moon.

page 128, In 1883, Baron Pierre de Coubertin of France, after visiting a rugby school in England: Wiles, K. 2017. "The First Modern Olympic Games." History Today. https://www.historytoday.com/archive/months-past/first-modern-olympic-games.

page 128, Originally held in Olympia, Greece, starting in the eighth century BC: "The Games." 2020. Penn Museum. https://www.penn.museum/sites/olympics/olymporigins.shtml.

page 128, The first modern Olympics took place in Greece in 1896: Wiles, "The First Modern Olympic Games," 2017.

page 128, "Miracle on Ice": "US Ice Hockey Rookies Conjure Up a Miracle on Ice." 1980. International Olympic Committee. https://www.olympic.org/news/us-ice-hockey-rookies-conjure-up-a-miracle-on-ice.

page 129, South African rugby game on June 14, 1994: "The Early History of Rugby in South Africa." 2019. South African History Online. https://www.sahistory.org.za/article/early-history-rugby-south-africa.

page 130, Best predictor of how healthy and happy a person is at eighty years old: Mineo, L. 2017. "Good Genes Are Nice, but Joy Is Better." *Harvard Gazette*. https://news.harvard.edu/gazette/story/2017/04/over-nearly-80-years-harvard-study-has-been-showing-how-to-live-a-healthy-and-happy-life/.

GRIEVING RITUALS

page 134, Garrano horses in northern Portugal: Mendonca, R. S., M. Ringhofer, P. Pinto, S. Inoue, and S. Hirata. 2020. "Feral Horses' (*Equus ferus caballus*) Behavior toward Dying and Dead Conspecifics." *Primates* 61 (1): 49–54.

page 134, Similar grief-like behaviors in horses: Dickinson, G. E., and H. C. Hoffmann. 2016. "The Difference between Dead and Away: An Exploratory Study of Behavior Change during Companion Animal Euthanasia." *Journal of Veterinary Behavior* 15: 61–65.

page 134, Thanatology, the study of death and the psychological and social conditions surrounding death: Anderson, J. R. 2016. "Comparative Thanatology." *Current Biology* 26 (13): R553–R556.

page 135, Mourning behavior can be costly, both physically and psychologically: King, B. J. 2013. "When Animals Mourn." *Scientific American* 309 (1): 62–67.

page 135, *How Animals Grieve*: King, B. J. 2013. *How Animals Grieve*. Chicago: University of Chicago Press.

page 135, Suite of behaviors called "mortuary": Anderson, "Comparative Thanatology," 2016.

page 135, Escalated behaviors that involve some measure of emotional distress, or grief, called "funerary" behaviors: ibid.

page 135, Two criteria to determine whether a nonhuman animal's response to death is considered grief: King, "When Animals Mourn," 2013.

page 136, Chimpanzees probably grieve in the human sense of the word: Sapolsky, R. M. 2016. "Psychiatric Distress in Animals versus Animal Models of Psychiatric Distress." *Nature Neuroscience* 19 (11): 1387–1389.

page 136, In Jane Goodall's chimpanzee study in Gombe: "The 'F' Family." 2017. Jane Goodall Institute. https://www.janegoodall.org.au/2017/03/the-f-family/.

page 136, Wild female chacma baboons display signs of grief at the loss: Engh, A. L., J. C. Beehner, T. J. Bergman, P. L. Whitten, R. R. Hoffmeier, R. M. Seyfarth, and D. L. Cheney. 2006. "Behavioural and Hormonal Responses to Predation in Female Chacma Baboons (*Papio hamadryas ursinus*)." *Proceedings of the Royal Society B: Biological Sciences* 273 (1587): 707–712.

page 136, Cortisol levels also go up in humans during bereavement: Alderton, D. 2011. *Animal Grief: How Animals Mourn*. Dorset, UK: Hubble & Hattie.

page 136, By extending social networks during mourning, cortisol levels and stress go down: Engh et al., "Behavioural and Hormonal Responses to Predation in Female Chacma Baboons," 2006.

page 137, Baboons also rely on an extended community to soothe the loss of a family member: ibid.

page 137, Social insects—such as ants, bees, and termites—also exhibit mortuary behaviors: Anderson, "Comparative Thanatology," 2016.

page 137, Sharks can also detect necromones in other sharks: Stroud, E. M., C. P. O'Connell, P. H. Rice, N. H. Snow, B. B. Barnes, M. R. Elshaer, and J. E. Hanson. 2014. "Chemical Shark Repellent: Myth or Fact? The Effect of a Shark Necromone on Shark Feeding Behavior." *Ocean and Coastal Management* 97: 50–57.

page 137, Even humans have putrefaction volatiles that act as necromones: Izquierdo, C., J. C. Gomez-Tamayo, J. C. Nebel, L. Pardo, and A. Gonzalez. 2018. "Identifying Human Diamine Sensors for Death Related Putrescine and Cadaverine Molecules." *PLOS Computational Biology* 14 (1): e1005945.

page 137, Signals that trigger a dog's ability to identify diseases in humans: Dickinson and Hoffmann, "The Difference between Dead and Away," 2016.

page 137, In one giraffe family in Kenya, researchers observed that a mother gave birth to a baby with a deformed foot: King, "When Animals Mourn," 2013.

page 138, The mother-offspring bond after the death of a calf is even stronger if the calf lived long enough to suckle: Bercovitch, F. B. 2020. "A Comparative Perspective on the Evolution of Mammalian Reactions to Dead Conspecifics." *Primates* 61 (1): 21–28.

page 138, In Jim and Jamie Dutcher's book: Jim and Jamie Dutcher. 2018. *The Wisdom of Wolves.* Washington, DC: National Geographic.

page 138, Carrying behavior has been documented in a number of species, including apes, monkeys, dolphins, and dingoes: Anderson, "Comparative Thanatology," 2016.

page 140, In the case of dolphins, mothers have been seen carrying their dead: King, "When Animals Mourn," 2013.

page 140, Chimp population in Guinea, an experienced mother carried the mummified remains of her baby for almost seventy days: Biro, D., T. Humle, K. Koops, C. Sousa, M. Hayashi, and T. Matsuzawa. 2010. "Chimpanzee Mothers at Bossou, Guinea Carry the Mummified Remains of Their Dead Infants." *Current Biology* 20 (8): R351–R352.

page 140, Why carrying the dead: Watson, C. F. I., and T. Matsuzawa. 2018. "Behaviour of Nonhuman Primate Mothers toward Their Dead Infants: Uncovering Mechanisms." *Philosophical Transactions of the Royal Society of London. Series B, Biological Sciences* 373 (1754): 20170261.

page 140, Carry, hold, or remain with the dead may have important physical and psychological benefits: Kingdon, C., E. O'Donnell, J. Givens, and M. Turner. 2015. "The Role of Healthcare Professionals in Encouraging Parents to See and Hold Their Stillborn Baby: A Meta-Synthesis of Qualitative Studies." *PLOS ONE* 10 (7): e0130059.

page 140, A recent human meta-study combining twelve studies from six countries: ibid.

page 142, Elephants have occasionally been observed returning: Goldenberg, S. Z., and G. Wittemyer. 2020. "Elephant Behavior toward the Dead: A Review and Insights from Field Observations." *Primates* 61 (1): 119–128.

page 142, Elephants spend visiting the remains of another elephant indicates more than mere curiosity: ibid.

notes

page 143, Elephants burying dead elephants by sprinkling dirt: Siegal, R. K. 1980. "The Psychology of Life after Death." *American Psychologist* 35 (10): 911–931.

page 144, Chimpanzees bury dead family members: ibid.

page 144, Many social animals have pragmatic mortuary rituals: Anderson, "Comparative Thanatology," 2016.

page 144, Early humans took burial further: Pettitt, P. 2018. "Hominin Evolutionary Thanatology from the Mortuary to Funerary Realm: The Palaeoanthropological Bridge between Chemistry and Culture." *Philosophical Transactions of the Royal Society B: Biological Sciences* 373 (1754): 20180212.

page 144, Paleolithic archeologists find the first evidence of funerary caching: ibid.

page 144, People began setting aside particular landscapes for the dead: ibid.

page 144, First evidence of an understanding of an afterlife: ibid.

page 144, Human culture and cooperation blossomed during the Pleistocene era: ibid.

page 145, Many people commemorate the dead in ways that reflect a belief in the existence of a soul: Siegal, "The Psychology of Life after Death," 1980.

page 145, Grief serves as a time to reflect: King, *How Animals Grieve*, 2013.

page 145, Group grieving just after a death is also an important steppingstone: Walsh, F., and M. McGoldrick, eds. 2004. *Living beyond Loss: Death in the Family*. 2nd ed. New York: W. W. Norton and Company.

page 146, The practice of hiring professional mourners, called "moirologists": "Professional Mourning." 2020. Wikipedia. https://en.wikipedia.org/wiki/Professional_mourning.

page 146, Professionals can be hired to attend wakes and engage in grieving, lamenting, even eulogizing, as well as providing comfort to the grieving family: ibid.

page 146, "Death doulas" serve as facilitators and grief counselors to the dying and their families: Watt, C. S. 2019. "End-of-Life Doulas: The Professionals Who Guide the Dying." *Guardian*. https://www.theguardian.com/lifeandstyle/2019/nov/06/end-of-life-doulas-the-professionals-who-help-you-die.

page 146, Mourning rituals serve to support and contain strong emotions: Walsh, F., and McGoldrick, M. 1991. "Loss and the Family: A Systemic Perspective." In *Living beyond Loss: Death in the Family*. W. W. Norton and Company.

page 146, Mourners come together with the grieving family for a short time: ibid.

page 146, Holding a wake, a funeral, and a burial, and then sharing a meal afterward: ibid.

page 146, Prevent the bereaved from experiencing a period of isolation surrounding their loss: ibid.

page 146, Serve as moments to remember the loss and help process grief: ibid.

page 147, Persistent symptoms of grief are at higher risk of getting cancer, heart disease, and hypertension: Iliya, Y. A. 2015. "Music Therapy as Grief Therapy for Adults with Mental Illness and Complicated Grief: A Pilot Study." *Death Studies* 39 (1–5): 173–184.

page 147, Grieving with an extended community helps to reduce stress and allows us to mature: Walsh and McGoldrick, "Loss and the Family: A Systemic Perspective," 1991.

page 147, Group grieving is important, psychologically and physiologically: Cheney, D. L., and R. M. Seyfarth. 2009. "Chapter 1: Stress and Coping Mechanisms in Female Primates." *Advances in the Study of Behavior* 39: 1–44.

page 147, This includes societal grief: King, "When Animals Mourn," 2013.

page 147, Grief for the loss of strangers evolved as survival: ibid.

page 147, Prolonged interactions with the dead: ibid.

page 148, Day of the Dead is a multiday celebration of the lives: Greenleigh, J., and R. R. Beimler. 1998. *The Days of the Dead: Mexico's Festival of Communion with the Departed*. Rohnert Park, CA: Pomegranate.

page 148, Tomb-sweeping: Zhang, L. 2007. "On the Custom of Tomb-Sweeping and Ancestor Worship in Yuanzaju." *Journal of Chongqing University (Social Science Edition)*. 2007-01.

page 148, All Souls' Day: USAG Benelux Public Affairs. 2017. "All Saints' Day Honors the Deceased." https://www.army.mil/article/196239/all_saints_day_honors_the_deceased.

page 149, Very effective therapy for grief: Iliya, "Music Therapy as Grief Therapy for Adults with Mental Illness and Complicated Grief," 2015.

RITUALS OF RENEWAL

page 153, Whales arrive in the tropical waters of Hawaii between the months of January and March: Craig, A. S., L. M. Herman, C. M. Gabriele, and A. A. Pack. 2003. "Migratory Timing of Humpback Whales (*Megaptera novaeangliae*) in the Central North Pacific Varies with Age, Sex and Reproductive Status." *Behaviour* 140 (8/9): 981–1001.

page 153, Birds use the length of day as an environmental cue to track seasonal changes: Dawson, A. 2007. "Seasonality in a Temperate Zone Bird Can Be Entrained by Near Equatorial Photoperiods." *Proceedings of the Royal Society B: Biological Sciences* 274 (1610): 721–725.

page 153, (Japanese quail) hormones will adjust within hours of arrival: Ball, G. F., and J. Balthazart. 2010. "Japanese Quail as a Model System for Studying the Neuroendocrine Control of Reproductive and Social Behaviors." *ILAR Journal* 51 (4): 310–325.

page 153, The rufous-winged sparrow of the Sonoran Desert needs a different internal clock: Brashears, A. 2012. "Singing in the Rain." Arizona State University School of Life Sciences. Ask a Biologist. https://askabiologist.asu.edu/explore/animals-seasons.

page 154, A bear needs to eat enough during the fall in order to survive a hibernation-like state: "The Brown Bear: Torpor or Hibernation?" 2017. Bear Sanctuary Domazhyr. https://www.bearsanctuary-domazhyr.org/our-bears/about-bears/brown-bear-torpor-or-hibernation.

page 154, Bears go through a physiological change at this time, called hyperphagia: "When Bears Prepare for Winter." 2018. National Park Service. https://www.nps.gov/articles/bears-winter.htm.

page 154, Salmon themselves follow environmental and hormonal cues that regulate their whole lives: Cooke, S. J., G. T. Crossin, and S. G. Hinch. 2011. "Pacific Salmon Migration: Completing the Cycle." In *Encyclopedia of Fish Physiology: From Genome to Environment*, edited by A. P. Farrell, 1945–1952. San Diego: Academic Press.

page 154, A monarch caterpillar metamorphoses into a butterfly and flies south and west: "Monarch Butterfly Migration and Overwintering." U.S. Forest Service. https://www.fs.fed.us/wildflowers/pollinators/Monarch_Butterfly/migration/index.shtml.

page 154, Seals molt in the spring to shed their winter coat for summer: Ling, J. K. 1970. "Pelage and Molting in Wild Mammals with Special Reference to Aquatic Forms." *Quarterly Review of Biology* 45 (1): 16–54.

notes

page 155, Elk and moose males grow enormous antlers: Langley, L. 2018. "Why Do Moose Shed Their Antlers?" *National Geographic.* https://www.nationalgeographic.com/news/2018/01/animals-antlers-moose-seasons-mating/.

page 155, Termites have a nuptial flight: van Huis, A. 2017. "Cultural Significance of Termites in Sub-Saharan Africa." *Journal of Ethnobiology and Ethnomedicine* 13 (8): 1–12.

page 155, Human biorhythms are affected by the same daylength and light-exposure factors: Brambilla, C., C. Gavinelli, D. Delmonte, M. C. Fulgosi, B. Barbini, C. Colombo, and E. Smeraldi. 2012. "Seasonality and Sleep: A Clinical Study on Euthymic Mood Disorder Patients." *Depression Research and Treatment* 2012: 978962.

page 155, Just being in nature, studies have shown, reduces stress levels, lowers blood pressure, decreases muscle tension, and lessens negative emotions like anxiety and depression: Park, B. J., Y. Tsunetsugu, T. Kasetani, T. Kagawa, and Y. Miyazaki. 2010. "The Physiological Effects of Shinrin-yoku (Taking in the Forest Atmosphere or Forest Bathing): Evidence from Field Experiments in 24 Forests across Japan." *Environmental Health and Preventive Medicine* 15 (1): 18–26.

page 155, Being aware of the cycles of nature—is beneficial: Crimmins, T. 2020. "To Ease Climate Anxiety, Reconnect with the Rhythms of the Seasons." *Scientific American.* https://blogs.scientificamerican.com/observations/to-ease-climate-anxiety-reconnect-with-the-rhythms-of-the-seasons/.

page 155, The mood-boosting aspect of spring is caused by the onset of warmer weather: Willis, J. 2015. "The Science of Spring: How a Change of Seasons Can Boost Classroom Learning." *Guardian.* https://www.theguardian.com/teacher-network/2015/apr/02/science-spring-how-seasons-classroom-learning.

page 155, Growth hormones even kick in during spring: Tendler, A., A. Bar, N. Mendelsohn-Cohen, O. Karin, Y. Korem, L. Maimon, T. Milo, et al. 2020. "Human Hormone Seasonality." bioRxiv preprint.

page 156, Many rituals surrounding the onset of spring begin on the spring equinox: "The Seasons, the Equinox, and the Solstices." Weather. https://www.weather.gov/cle/seasons.

page 156, Rituals surrounding the spring or vernal equinox have been around for centuries: "Spring Equinox." BBC. https://www.bbc.co.uk/religion/religions/paganism/holydays/springequinox.shtml.

page 156, Ostara: ibid.

page 156, Christian Easter and Jewish Passover are observed at this time as well: "Ostara Facts and Worksheets." 2019. Kids Konnect. https://kidskonnect.com/holidays-seasons/ostara/.

page 156, Nowruz: ibid.

page 156, In Roman mythology, the god Mithras was resurrected on the spring equinox: ibid.

page 156, Mayans performed rituals at the temple of Kukulcan: ibid.

page 156, Ostara rituals, both ancient and currently practiced . . . include setting up altars: ibid.

page 157, The summer and winter solstices occur when the sun is at its greatest distance from the celestial equator: "The Seasons, the Equinox, and the Solstices."

page 157, Stonehenge is aligned like the Mayan structures: Greenspan, R. E. 2019. "Here's Why Stonehenge Is Connected to the Summer Solstice." *Time.* https://time.com/5608296/summer-solstice-stonehenge-history/.

page 157, This reduced light exposure is the cause of seasonal affective disorder: Rosenthal, N. E., D. A. Sack, C. Gillin, A. J. Lewy, F. K. Goodwin, Y. Davenport, P. S. Mueller, D. A. Newsome, and T. A. Wehr. 1984. "Seasonal Affective Disorder." *Archives of General Psychiatry* 41: 72–80.

page 157, Activities and rituals that help us acknowledge seasonal changes are helpful ways to reconnect us to nature: Crimmins, "To Ease Climate Anxiety, Reconnect with the Rhythms of the Seasons," 2020.

page 157, Spending time outdoors can help connect us to the seasons: Sorin, F. 2015. "13 Reasons Why Gardening Is Good for Your Health." Gardening Gone Wild. https://gardeninggonewild.com/13-reasons-why-gardening-is-good-for-your-health/.

page 157, A record of the first cherry blossoms in Kyoto goes back to the ninth century: Aono, Y., and K. Kazui. 2008. "Phenological Data Series of Cherry Tree Flowering in Kyoto, Japan, and Its Application to Reconstruction of Springtime Temperatures since the 9th Century." *International Journal of Climatology* 28 (7): 905–914.

page 158, Dusky-footed wood rats, more commonly known as "packrats," are known to place bay leaves in their nest as a fumigant to reduce parasites: Hemmes, R. B., A. Alvarado, and B. L. Hart. 2002. "Use of California Bay Foliage by Wood Rats for Possible Fumigation of Nest-Borne Ectoparasites." *Behavioral Ecology* 13 (3): 381–385.

page 158, The nocturnal beach mouse does a bit of spring cleaning of its own: Gentry, J. B., and M. H. Smith. 1968. "Food Habits and Burrow Associates of *Peromyscus polionotus*." *Journal of Mammalogy* 49 (3): 562–565.

page 158, The house mouse engages in the same behavior during March and April: Schmid-Holmes, S., L. C. Drickamer, A. S. Robinson, and L. L. Gillie. 2001. "Burrows and Burrow-Cleaning Behavior of House Mice (*Mus musculus domesticus*)." *American Midland Naturalist* 146 (1): 53–62.

page 158, Birds that reuse nest sites, like starlings, also "clean house": Mazgajski, T. D. 2019. "Nest Site Preparation and Reproductive Output of the European Starling (*Sturnus vulgaris*)." *Avian Biology Research* 6 (2): 119–126.

page 158, The honeybee hive is cleanest and most sterile environments in nature: Simone, M., J. D. Evans, and M. Spivak. 2009. "Resin Collection and Social Immunity in Honeybees." *Evolution* 63 (11): 3016–3022.

page 158, Many cases of nest-cleaning in social insects: Bot, A. N. M., C. R. Currie, A. G. Hart, and J. J. Boomsma. 2001. "Waste Management in Leaf-Cutting Ants." *Ethology Ecology and Evolution* 13: 225–237.

page 159, Cleaning stations are well known among fish communities, such as cleaner wrasse: Bshary, R. 2003. "The Cleaner Wrasse, *Labroides dimidiatus*, Is a Key Organism for Reef Fish Diversity at Ras Mohammed National Park, Egypt." *Journal of Animal Ecology* 72 (1): 169–176.

page 159, Origin of spring cleaning in humans stems from the Persian New Year: Thomas, L. 2014. "2. Cleaning as a Cultural Impulse." In *Why Cleaning Has Meaning: Bringing Wellbeing into Your Home*. Edinburgh, UK: Floris Books.

page 159, Spring-cleaning ritual is attributed to the ancient Jewish practice of thoroughly cleansing the home in anticipation of the weeklong celebration of Passover: ibid.

page 159, Catholics have a similar spring-cleaning ritual leading up to Lent: ibid.

page 159, Spring cleaning is especially common in environments that have cold winters: Thomas, "2. Cleaning as a Cultural Impulse," 2014.

notes

page 160, Cleaning is important to help us maintain our physical and mental health: McDonnell, J. "The Health Benefits of Cleaning." Rush University Medical Center. https://www.rush.edu/health-wellness/discover-health/health-benefits-cleaning.

page 160, Fatty fish and dark leafy greens, vegetable oils, complex carbohydrates, and berries help shape our cognitive health: Mosconi, L. 2018. "Food for Thought: The Smart Way to Better Brain Health." *Guardian*. https://www.theguardian.com/lifeandstyle/2018/oct/13/food-diet-what-you-eat-affects-brain-health-dementia.

page 161, By 2030 half of all adults within the United States will suffer from obesity: Ward, Z. J., S. N. Bleich, A. L. Cradock, J. L. Barrett, C. M. Giles, C. Flax, M. W. Long, and S. L. Gortmaker. 2019. "Projected U.S. State-Level Prevalence of Adult Obesity and Severe Obesity." *New England Journal of Medicine* 381 (25): 2440–2450.

page 161, People have more body fat today than they did twenty years ago: Brown, R. E., A. M. Sharma, C. L. Ardern, P. Mirdamadi, P. Mirdamadi, and J. L. Kuk. 2016. "Secular Differences in the Association between Caloric Intake, Macronutrient Intake, and Physical Activity with Obesity." *Obesity Research and Clinical Practice* 10 (3): 243–255.

page 161, The overuse of antibiotics: Zhang, S., and D. C. Chen. 2019. "Facing a New Challenge: The Adverse Effects of Antibiotics on Gut Microbiota and Host Immunity." *Chinese Medical Journal* 132 (10): 1135–1138.

page 161, Elephants already knew this and seed their own gut by eating the dung of other family members: Soave, O., and C. D. Brand. 1991. "Coprophagy in Animals: A Review." *Cornell Veterinarian* 81 (4): 357–364.

page 161, Change in the makeup of grass communities: Li, G., B. Yin, J. Li, et al. 2020. "Host-Microbiota Interaction Helps to Explain the Bottom-Up Effects of Climate Change on a Small Rodent Species." *Multidisciplinary Journal of Microbial Ecology* 14: 1795–1808.

page 161, A flair up of dangerous bacteria within the gut flora of the endangered saiga antelope: Kock, R. A., M. Orynbayev, S. Robinson, S. Zuther, N. J. Singh, W. Beauvais, E. R. Morgan, et al. 2018. "Saigas on the Brink: Multidisciplinary Analysis of the Factors Influencing Mass Mortality Events." *Science Advances* 4 (1): eaao2314.

page 163, As little as eleven hours of meditation over time: Tang, Y. Y., Q. Lu, X. Geng, E. A. Stein, Y. Yang, and M. I. Posner. 2010. "Short-Term Meditation Induces White Matter Changes in the Anterior Cingulate." *Proceedings of the Natural Academy of Sciences USA* 107 (35): 15649–15652.

RITUALS OF TRAVEL & MIGRATION

page 167, Wildebeest and zebra migrations occurred within Botswana: Lindsey, P. A., C. L. Masterson, A. L. Beck, and S. Romañach. 2012. "Ecological, Social and Financial Issues Related to Fencing as a Conservation Tool in Africa." In *Fencing for Conservation: Restriction of Evolutionary Potential or a Riposte to Threatening Processes?*, edited by M. J. Somers and M. W. Hayward, 215–234. New York: Springer.

page 167, There are efforts to reconnect a safe passage: Chase, M. J., and C. R. Griffin. 2009. "Elephants Caught in the Middle: Impacts of War, Fences and People on Elephant Distribution and Abundance in the Caprivi Strip, Namibia." *African Journal of Ecology* 47: 223–233.

page 167, Over two million wildebeest and two hundred thousand zebra and small antelope migrate: Harris, G., S. Thirgood, J. G. C. Hopcraft, J. P. G. M. Cromsight, and J. Berger. 2009.

"Global Decline in Aggregated Migrations of Large Terrestrial Mammals." *Endangered Species Research* 7: 55–76.

page 167, The snow goose, for example, migrates south to the United States and Mexico to escape cold: Abraham, K. F., R. L. Jefferies, and R. T. Alisauskas. 2005. "The Dynamics of Landscape Change and Snow Geese in Mid-continent North America." *Global Change Biology* 11 (6): 841–855.

page 167, Humans once pursued herds of American buffalo or woolly mammoths: "The Earliest Humans in Yellowstone." 2019. National Park Service. https://www.nps.gov/yell/learn/historyculture/earliest-humans.htm.

page 167, The eastern gray whale migrates from Russian waters to Mexico and back: Mate, B. R., V. Y. Ilyashenko, A. L. Bradford, V. V. Vertyankin, G. A. Tsidulko, V. V. Rozhnov, and L. M. Irvine. 2015. "Critically Endangered Western Gray Whales Migrate to the Eastern North Pacific." *Biology Letters* 11 (4): 20150071.

page 167, Tiny arctic tern makes the longest migration of any animal in the world: Egevang, C., I. J. Stenhouse, R. A. Phillips, A. Petersen, J. W. Fox, and J. R. Silk. 2010. "Tracking of Arctic Terns *Sterna paradisaea* Reveals Longest Animal Migration." *Proceedings of the Natural Academy of Sciences USA* 107 (5): 2078–2081.

page 168, Humans slowly shifted away from their hunter-gatherer and migratory behavior during the Pleistocene era: Marlowe, F. W. 2005. "Hunter-Gatherers and Human Evolution." *Evolutionary Anthropology: Issues, News, and Reviews* 14 (2): 54–67.

page 168, Overhunting and extinction of major prey items—like woolly mammoths, mastodons, and giant sloths: Sandom, C., S. Faurby, B. Sandel, and J. Svenning. 2014. "Global Late Quaternary Megafauna Extinctions Linked to Humans, Not Climate Change." *Proceedings of the Royal Society B: Biological Sciences* 281 (1787): 20133254.

page 168, Humans migrate mostly for socio-economic or political reasons: Hagen-Zanker, J. 2008. "Why Do People Migrate? A Review of the Theoretical Literature." Maastrcht Graduate School of Governance Working Paper No. 2008/WP002.

page 168, Ashura procession in Copenhagen: Pedersen, M. H., and M. Rytter. 2017. "Rituals of Migration: An Introduction." *Journal of Ethnic and Migration Studies* 44 (16): 2603–2616.

page 168, Muslims traveling to the Saudi Arabian city of Mecca, the holy city of Islam: Zeidan, A. 2020. "Hajj." Britannica. https://www.britannica.com/topic/hajj.

page 168, Roman Catholics pilgrimage to Rome: "Pilgrimage: Rome." BBC. https://www.bbc.co.uk/bitesize/guides/z84dtfr/revision/6.

page 168, Indigenous Australians believe songlines, or "dream tracks,": Somerville et al., "Walking Contemporary Indigenous Songlines as Public Pedagogies of Country," 2019.

page 169, Vision quests are another type of sacred ritual travel: Krown, M. K. 2009. "Huffington Post: What Is a Vision Quest and Why Do One?" School of Lost Borders. http://www.schooloflostborders.org/content/huffington-post-what-vision-quest-and-why-do-one.

page 169, Travel . . . healthy it is, both physically and psychologically: Chen, C., and J. F. Petrick. 2013. "Health and Wellness Benefits of Travel Experiences." *Journal of Travel Research* 52 (6): 709–719.

page 171, Some plant species need fire for their seeds to germinate. This is particularly true of chaparral ecosystems in California, where hot fires, every few decades, stimulate new plant growth: Barro, S. C., and S. G. Conard. 1991. "Fire Effects on California Chaparral Systems: An Overview." *Environment International* 17 (2–3): 135–149.

notes

page 171, Early humans developed a relationship with fire that shaped our evolution: Hebrew University of Jerusalem. 2008. "Fire Out of Africa: A Key to the Migration of Prehistoric Humans." ScienceDaily. www.sciencedaily.com/releases/2008/10/081027082314.htm.

page 171, Modern human activities have increased the frequency and severity of fire: Bowman, D. M., J. Balch, P. Artaxo, W. J. Bond, M. A. Cochrane, C. M. D'Antonio, R. Defries, et al. 2011. "The Human Dimension of Fire Regimes on Earth." *Journal of Biogeography* 38 (12): 2223–2236.

page 171, The Cassin's vireo and the Swainson's thrush, declined after wildfires: Smucker, K. M., R. L. Hutto, and B. M. Steele. 2005. "Changes in Bird Abundance after Wildfire: Importance of Fire Severity and Time since Fire." *Ecological Applications* 15 (5): 1535–1549.

page 176, The Japanese call this experience "forest bathing"—or *shinrin-yoku*: Smith, C. 2014. "Forest Bathing." *Psychology Today*. https://www.psychologytoday.com/us/blog/shift/201409/forest-bathing.

page 176, Forest bathing was shown to reduce stress levels, lower pulse rate and blood pressure: Park et al., "The Physiological Effects of Shinrin-Yoku," 2010.

page 179, The gray wolf was reintroduced to Yellowstone in the mid-1990s, and since then, it has had a very positive impact on the environment: "Wolf Restoration." 2020. National Park Service. https://www.nps.gov/yell/learn/nature/wolf-restoration.htm.

page 180, Journeys in nature, and even travel itself, are restorative: Berman, M. G., J. Jonides, and S. Kaplan. 2008. "The Cognitive Benefits of Interacting with Nature." *Psychological Sciences* 19: 1207–1212.

page 180, The travel bug is encoded in our genes: Crouch, G. I. 2013. "*Homo sapiens* on Vacation." *Journal of Travel Research* 52 (5): 575–590.

page 180, Our ancestors were migratory hunter-gatherers: Marlowe, "Hunter-Gatherers and Human Evolution," 2005.

page 181, Even planning a trip makes us happier: Nawijn, J., M. A. Marchand, R. Veenhoven, and A. J. Vingerhoets. 2010. "Vacationers Happier, but Most Not Happier after a Holiday." *Applied Research in Quality of Life* 5 (1): 35–47.

page 181, A trip can improve blood pressure: Hruska, B., S. D. Pressman, K. Bendinskas, and B. B. Gump. 2020. "Vacation Frequency Is Associated with Metabolic Syndrome and Symptoms." *Psychology and Health* 35 (1): 1–15.

page 181, Boost our immune system: Vinocur, L. 2015. "10 Reasons Why Vacations Matter." Take Back Your Time. https://www.takebackyourtime.org/why-vacations-matter/10-reasons-to-vacation/.

page 181, Experiencing new surroundings offers us a fresh perspective on life, as well as a new appreciation for the place we call home: Rowan Kelleher, S. 2019. "This Is Your Brain on Travel." *Forbes*. https://www.forbes.com/sites/suzannerowankelleher/2019/07/28/this-is-your-brain-on-travel/#77b646db2be6.

page 181, Students who studied abroad were 20 percent more likely to succeed: Darian, N. 2015. "The 10 Lessons to Learn from Traveling." *A Leading Study Abroad Blog* (blog), *HuffPost*. https://www.huffpost.com/entry/the-10-lessons-to-learn-f_b_8056918.

page 183, In *Everything in Its Place*, Oliver Sacks writes: Sacks, O. 2019. *Everything in Its Place: First Loves and Last Tales*. New York: Knopf.

page 184, In Buddhism, compassion is the result of knowing one is part of a greater whole: Goetz, J. 2004. "Research on Buddhist Conceptions of Compassion: An Annotated Bibliography." *Greater Good*.

ACKNOWLEDGMENTS

THE IDEA FOR this book was inspired by attending SUMMIT LA, an annual ideas festival held in Los Angeles. I am grateful to my close friend Kerry Gilmartin for inviting me to join her at this four-day event as it wasn't the kind of thing that would have fallen onto my radar. We were both in the middle of a life transition and were buoyed by the diverse attendees and innovative venue.

I was also in the process of undergoing a career transition back to my long-term elephant hearing research and away from the biotech industry where I was a principal investigator on several NASA projects. At the time, I was conflicted about how much is being invested in the space industry when our planet and its inhabitants are in desperate need of our resources and attention. There is no planet B that we know of, and while it is amazing that a few individuals have been able to invest their personal wealth in searching for habitable planets, our own very unique planet is in crisis. I was glad to be able to turn my full attention back to elephants.

This transition triggered further reflection, and a number of sessions at the festival provided food for thought that led to the kernel of an idea for a book about the striking similarities between rituals performed throughout the animal kingdom.

In conversations with another dear friend and agent, Tina Seelig, my ideas were quickly shaped into a book proposal, and in short order she found a home for the book with Chronicle Prism. I can't thank her enough for forging this relationship.

In a first conversation with my editor, Eva Avery, I knew immediately that I was in the right hands—down to a shared connection with Hawaii. She traced her own deep passion for the natural world to having spent her childhood in Hilo on the Big Island. This book is truly a product of a writer-editor match made in heaven. I am so grateful for Eva's passion and vision for this book as it took shape. Our many conversations and her confidence in the book and, ultimately, in me guided me to the finish line. I am humbled and most grateful. And I am also especially thankful to the cover artist, Pamela Geismar, and the design team; to Mark Tauber, Jennifer Jensen, and the production team: Beth Weber, Tera Killip, Cecilia Santini, and for the rest of the folks at Chronicle for their enthusiasm for this book.

I'd also to thank the elephant staff at the Oakland Zoo for all the time they spent with me and Donna the elephant during her target training, especially Colleen, Gina, and Jeff Kinzley. And I'm most grateful to the Etosha Ecological Institute and the Ministry of Environment and Tourism of Namibia for allowing us to continue our long-term elephant research within Etosha National Park through our nonprofit, Utopia Scientific—the site and our experiences there provided the foundation for this book. I'm also grateful to Stanford University's VPUE grant program for supporting me and Stanford undergraduate students as research assistants for over a decade of fieldwork. I'd also like to thank the Falconwood Foundation for supporting both the coral reef research and the flamingo research covered in this book. I am grateful for grant support for our elephant field work from The Elephant Sanctuary and private donors to Utopia Scientific. Part of the work discussed in Spoken Rituals is ongoing work supported by NIDCD grant #5K01DC017812-01 at the Eaton Peabody Lab at Harvard Medical School.

I thank Vernon Presley for sharing his experiences of elephant deaths in captivity with me and for his compassion in understanding and accommodating the elephant's need to grieve. I'm also grateful to

acknowledgments

Lyn Miles for sharing her intimate stories about Chantek the "orangutan person" while raising him and teaching him sign, and for being an inspiring advocate of our great ape cousins. I am grateful to Jim and Jamie Dutcher for sharing their amazing experiences while studying the Sawtooth wolf pack and especially stories about Lakota.

I am indebted to Margaret French Isaac for inspiring me to transform my gut microbiome, which most likely has added years to my life and my husband's as well, and for her faith in my ability to persevere—and for all the friends and family who were later inspired to join me in doing the same; your success is my success.

I am grateful to another dear friend, Tanya Meyer, for her insightful read of the manuscript and for the many fun discussions of ritual thereafter. I'm also grateful to my sister-in-law, Anne O'Connell Obermeyer, for her early read of the Play and Grieving chapters and for her thoughtful suggestions.

I'd like to thank my husband, Tim Rodwell, for his patience in putting up with many late nights of writing and numerous discussions about ritual and our experiences in the wild that contributed to this book. Thank you for being such a great partner in seeking adventures in nature together as well as the perfect photograph.

I want to thank Jeff Campbell for his sensitivity and wisdom provided during his copyedit. I am extremely fortunate to have had such an enthusiastic student intern, Jodie Berezin, who helped me with the referencing of this book. I am especially grateful to my sister, Siobhan, for the countless hours spent in final stages of writing, helping me to clarify passages and draw more vivid imagery out of my head and onto the page. Only my sister could do that!

Lastly, I thank my parents, Dan and Aline O'Connell, for their continued encouragement of me and my writing and for their many reads of this manuscript even in its earliest stages. And for instilling in me a passion for nature, art, and travel from a young age.

ABOUT THE AUTHOR

DR. CAITLIN O'CONNELL is on the faculty at Harvard Medical School in the Eaton Peabody Lab at Massachusetts Eye & Ear, where she continues her decades-long journey to improve hearing for the hearing impaired with lessons learned from elephant research currently funded by the National Institutes of Health. She is also a Faculty Associate at Harvard's Center for the Environment. Previously, she was Director of Life Sciences at HNu Photonics for five years, where she was principle investigator on a NASA research project and experimental platform that launched to the International Space Station. Before taking the position at Harvard, she was Adjunct Professor in the Department of Otolaryngology, Head & Neck Surgery at Stanford University School of Medicine for more than a decade. At the same time, she is an award-winning author and photographer and has studied elephants for almost thirty years, having written dozens of scientific papers, numerous feature magazine articles, and two memoirs about her experiences, *The Elephant's Secret Sense* and *Elephant Don*. Her passion for science writing compelled her to develop a creative science writing class for Stanford and the *New York Times* that she taught for several years, while working on, and the focus of the award-winning Smithsonian documentary *Elephant King*.

Caitlin has authored eight popular books about elephants, including her award-nominated *Ivory Ghosts*, a thriller series about the ivory trade inspired by her experiences investigating the ivory trade, which she converted into a graphic novel and screenplay. All of her nonfiction books include photographs that she and her husband, Tim Rodwell, have taken at their field site and elsewhere. She is published by such top houses as Houghton Mifflin, Simon & Schuster, and Random House. Their book collaborations have resulted in numerous awards, particularly the Sibert Honor and Horn Book Honor, along with four other awards for *The Elephant Scientist* (Houghton Mifflin Harcourt) in 2012. In 2014, *A Baby Elephant in the Wild* (Houghton Mifflin Harcourt) was awarded Outstanding Science Trade Book and chosen as a Junior Library Guild Selection.

Their award-winning photography and videography have appeared in *National Geographic* magazine and on National Geographic WILD, in *Smithsonian Magazine* and on the Smithsonian Channel, and in the *New York Times*, *National Wildlife* magazine, *Discover*, *Science News*, *Africa Geographic*, *Highlights*, and many other national and international magazines, scientific journals, newspapers, and online media. Caitlin's TEDYouth talk about elephants has almost, sixty thousand views. She has blogged from her elephant field site for both the *New York Times* and *National Geographic*.

Caitlin and Tim founded and run a nonprofit organization, Utopia Scientific (www.utopiascientific.org) focused on science messaging and education. They also co-founded an independent production company, Triple Helix Productions, with a mandate to develop more accurate and entertaining science content for the entertainment industry. Their home base is in San Diego, where Tim is an MD/PhD, MPH, and associate professor in the Department of Medicine at the University of California, San Diego.

You can find Caitlin's books at www.caitlineoconnell.com.